ADVANCES IN
MOLECULAR MODELING

Volume 3 • 1995

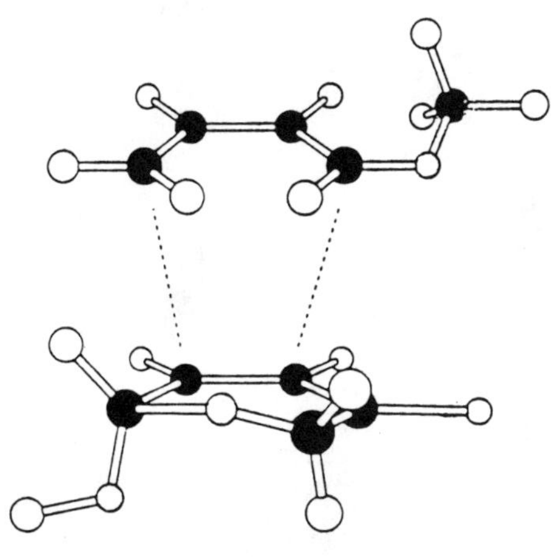

ADVANCES IN MOLECULAR MODELING

Editor: DENNIS LIOTTA

Department of Chemistry

Emory University

VOLUME 3 • 1995

Greenwich, Connecticut   London, England

CONTENTS

LIST OF CONTRIBUTORS

C. Webster Andrews   Burroughs Wellcome Company
Research Triangle Park
North Carolina

David W. Barry   Burroughs Wellcome Company
Research Triangle Park
North Carolina

D. Ross Boswell   Department of Chemistry
University of Canterbury
Christchurch, New Zealand

Edward E. Coxon   Department of Chemistry
University of Canterbury
Christchurch, New Zealand

James M. Coxon   Department of Chemistry
University of Canterbury
Christchurch, New Zealand

Phillip A. Furman   Burroughs Wellcome Company
Research Triangle Park
North Carolina

Deborah K. Jones   Department of Chemistry
Emory University
Atlanta, Georgia

Dennis Liotta   Department of Chemistry
Emory University
Atlanta, Georgia

D. Quentin McDonald

Department of Chemistry
University of Canterbury
Christchurch, New Zealand

George R. Painter

Burroughs Wellcome Company
Research Triangle Park
North Carolina

PREFACE

The field of molecular modeling continues to evolve rapidly and to grow in its diversity. Consistent with this growth, Volume 3 of this series presents a potpourri of articles dealing with several different aspects of this field. Some of the studies report on the use of modeling techniques to gain insight into the origins of enantioselectivity and diastereoselectivity in a number of important chemical reactions. Others deal with modeling methodology and the problems associated with obtaining accurate structural information about a ubiquitous class of compounds, carbohydrates. Finally, the creative merging of experimental and theoretical methods demonstrates the power modeling techniques for elucidating accurate, three-dimensional models of important biomacromolecules, such as the polymerase domain of HIV reverse transcriptase.

Although vastly different in scope, each of these studies shares a common theoretical foundation and each has the simple goal of providing insight which is complementary to that which can be obtained from experimental methods. I hope the reader will find all of these contributions to be interesting, informative, and thought-provoking.

Dennis Liotta
Editor

ORIGINS OF THE ENANTIOSELECTIVITY OBSERVED IN OXAZABOROLIDINE-CATALYZED REDUCTIONS OF KETONES

Dennis Liotta and Deborah K. Jones

Advances in Molecular Modeling
Volume 3, pages 1–19.
Copyright © 1995 by JAI Press Inc.
All rights of reproduction in any form reserved.
ISBN: 1-55938-326-7

I. INTRODUCTION

Reports describing the enantioselective reduction of ketones using stoichiometric borane in the presence of catalytic amounts of chiral oxazaborolidine represent a significant advance in rational reagent design.[1-6] In the 1980s, Itsuno et al.[1] investigated enantioselective ketone reductions using borane as the reductant in the presence of a chiral vicinal amino alcohol, (S)-2-amino-3-methyl-1,1-diphenyl-butan-1-ol (Figure 1).

Reductions using the corresponding alcohol derivative from (S)-leucine were also investigated. A high degree of success was achieved by the Itsuno group using these reduction processes. This is illustrated by the example show in Figure 1 (i.e., reduction of acetophenone to (R)-1-phenylethanol in 95%ee).

The Corey group took the Itsuno system shown in Figure 1 and studied it in great detail using ^{1}H and ^{11}B NMR to gain insight into the mechanism of this reduction.[2a] They found that an oxazaborolidine was formed between borane and the amino alcohol present and that this reagent subsequently behaved as the catalyst in the reduction (see Figure 2a). Further studies by Corey et al. showed that when (S)-(−)-2-(diphenylhydroxymethyl)pyrrolidine ((S)-diphenylprolinol) was heated with borane, this also formed an oxazaborolidine which had superior catalytic properties to the original system (Figure 2b). Excellent yields and enantioselectivities were obtained with this catalyst for a range of ketones with one large and one small substituent. The enantioselectivity was proposed to be due to steric constraints, forcing the larger substituent (R_L) to occupy the less hindered *exo* position (i.e., *anti* to the ring boron substituent) upon coordination of the ketone with the catalyst.

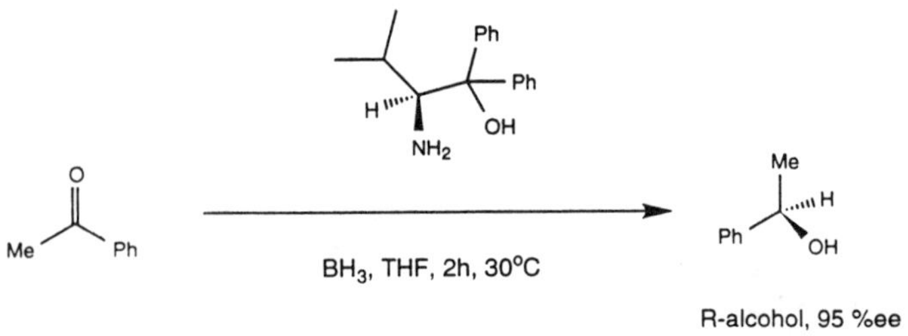

Figure 1. One of the systems studied by the Itsuno group.

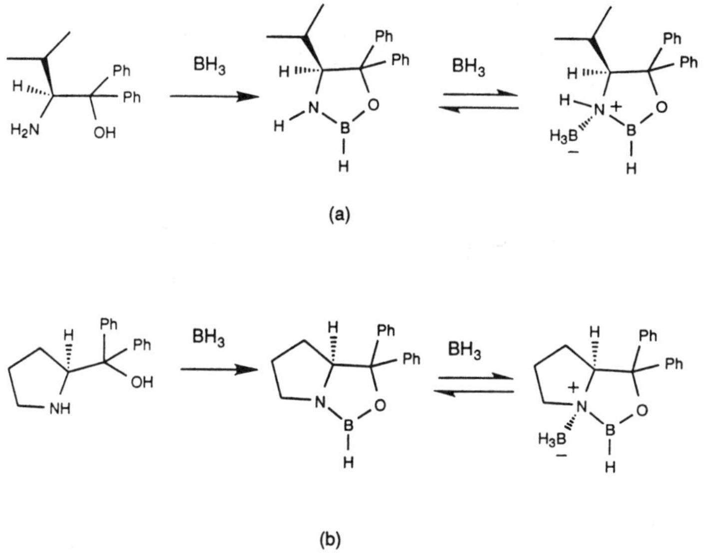

Figure 2. (a) (*S*)-2-amino-3-methyl-1,1-diphenylbutan-1-ol system. (**b**) (*S*)-(–)-2(diphenylhydroxymethyl)pyrrolidine system.

These enantioselective reductions are thought to occur by (a) initial complexation of borane to the ring nitrogen, (b) coordination of the ketone oxygen to the electrophilic ring boron, and, finally, (c) hydrogen transfer from the NBH_3^- unit to the carbonyl carbon via a six-membered cyclic transition state (Figure 3).

The Corey group have subsequently made many modifications to their original oxazaborolidine system (R_1 = phenyl and R_2 = H, see Figure 4). Studies have shown that the system which is air and moisture sensitive, the B-methylated system (R_1 = phenyl and R_2 = Me) is much easier to handle and in general gives slightly higher or the same enantioselectivity as observed with the original system.[2b] Catalysts in which the boron substituent is a *n*-butyl group have been shown to be of similar utility to the B-methylated compounds.[2j]

The nature of the R_1 substituents is of considerably greater significance than that of the boron substituent, R_2. The Corey group have found that when R_1 is either phenyl or β-napthyl, the resulting catalysts give good

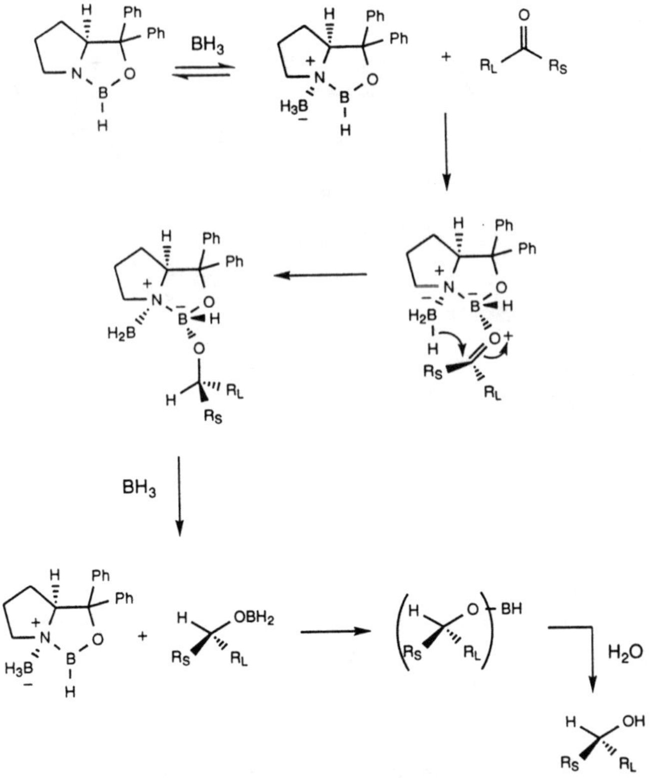

Figure 3. Proposed mechanism of reduction.

EFFECTIVE

INEFFECTIVE

R_1 = Ph, β-naphthyl

R_2 = H, Me, Bu

R_1 = α-naphthyl, H, i-Pr, o-MeOPh

Figure 4. Modifications made by the Corey group.

enantioselectivity. Conversely, α-napthyl, *i*-propyl, *o*-MeO-phenyl and H have all been shown to be ineffective substituents.[2i]

Dubbed the CBS process after the first initial of the authors second names on the original 1987 communication,[2a] this reduction has been shown by the Corey group to be of great synthetic utility. It has been successfully used in the synthesis of a number of natural products. Examples include the platelet activating factor antagonists, ginkgolide B[2c] and A,[2f] the potent adenylate cyclase activator forskolin,[2d] the β-adrenoreceptor agonist isoproterenol,[2k] and the serotonin-uptake inhibitor fluoxetine.[2g] Stereochemistry at the C(15) position in prostaglandin synthesis is simply and effectively controlled using this methodology.[2b]

Corey et al. have also investigated replacing borane with catecholborane. Using borane as the stoichiometric reductant often results in interference with functionality that is sensitive to borane (e.g., olefins and amides) as well as a loss of enantioselectivity at low temperatures. Using catecholborane allows low temperatures to be employed, resulting in enhanced selectivity and suppression of uncatalyzed reduction products.[2i,j]

Most of the oxazaborolidines used by Corey are derived from L-amino acids; use of the D-amino acid derived oxazaborolidine gives the other enantiomer of the secondary alcohol to that obtained from the L-amino acid derivative. The major disadvantage of this approach is that D-amino acids are expensive and not readily accessible. In an attempt to overcome this problem, Rao et al.[3a] developed a series of six-membered oxazaborolidines obtained from (*R*)- and (*S*)-α,α-diphenyl-2-piperidine methanol. The lower enantiomeric excesses observed using the six-membered oxazaborolidines (5–10% lower than to Corey's work) was assumed to be due to the decrease in steric influence in the case of six-membered compared to the five-membered oxazaborolidines. Rao[3b] confirmed this postulate with the synthesis of four-membered oxazaborolidines prepared from (*R*)- and (*S*)-α,α-diphenyl-2-azetidine methanols. Use of these catalysts gave enantiomeric excesses of 95 to 97%.

Despite the obvious synthetic interest in these oxazaborolidine catalysts, a clear-cut analysis of why such excellent enantioselectivities are observed has not been proposed. Ab initio calculations performed by Nevalainen[7] investigated the mechanism of reduction. A model system

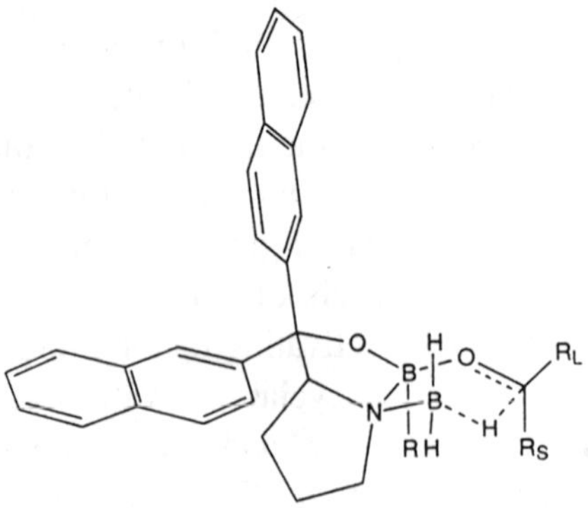

Figure 5. Representation of the boat transition state proposed by Corey et al.

was used as the basis of these calculations and hence the effect of varying substituents on the oxazaborolidine catalyst was not addressed.

Without the insight of thorough computational studies, proposals have been made about the nature of the reduction transition state. Both Corey et al.[2i] and Evans[8] have suggested that these processes occur by way of a boat transition state. Figure 5 is a representation of the proposed boat transition state.

II. METHODS

Our approach to studying these reactions was to use molecular orbital methods (MOPAC 5.0)[9] to investigate the origins of the selectivity of this reduction and the effects of substituent modifications. Such information should prove invaluable in the rational design of the next generation of oxazaborolidine catalysts.

The starting geometries for the complexes studied (Table 1) were generated using the molecular mechanics package PCMODEL,[10] with geometry optimization using the MMX force field.[11] Since parameters were not available for the boron functionality present in the oxazaborolidine system, the geometry was approximated using an sp^3 C type to mimic the tetrahedral boron. Subsequent geometry optimization using the MNDO Hamiltonian,[12] which has all the appropriate parameters to handle boron functionality, corrected for the original approximation made in the molecular mechanics calculations.

Table 1. Oxazaborolidine/Ketone/Borane Complexes Investigated Computationally

	R_1	R_2	R_L	R_S
1	Ph	H	Ph	Me
2	Ph	H	Ph	Et
3	Ph	H	^{t}Bu	Me
4	Ph	Me	Ph	Me
5	Ph	Me	Ph	Et
6	Ph	Me	^{t}Bu	Me
7	Ph	Et	Ph	Et
8	β-Naph	H	Ph	Me
9	^{i}Pr	Me	Ph	Me
10	α-Naph	H	Ph	Me
11	o-MeOPh	H	Ph	Me
12	o-MeOPh	Me	Ph	Me
13	H	H	Ph	Me
14	H	Me	Ph	Me

The reaction surface for the hydrogen transfer was examined for a variety of oxazaborolidine/ketone/borane complexes (see Table 1). The surfaces were computed by systematically varying two reaction coordinates in 0.1 Å increments: (1) the B–H bond of the hydrogen being transferred from the coordinated borane and (2) the C–H bond being formed between the transferring hydrogen and the carbonyl carbon being reduced (Figure 6). At each point on the surface, other than these two geometrical constraints, the structure was optimized.

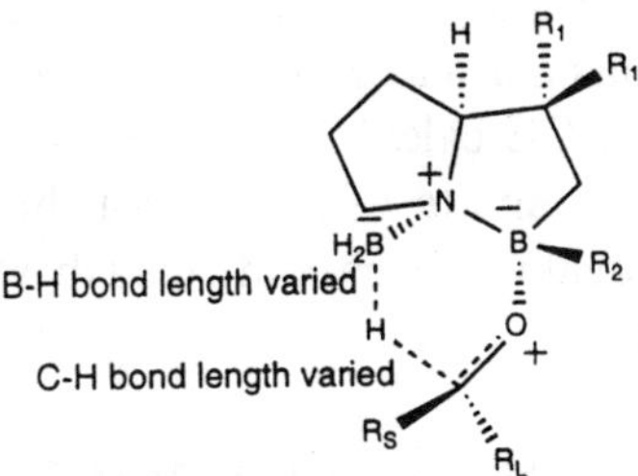

Figure 6. General oxazaborolidine/ketone/borane complex used to calculate the hydride transfer reaction surfaces.

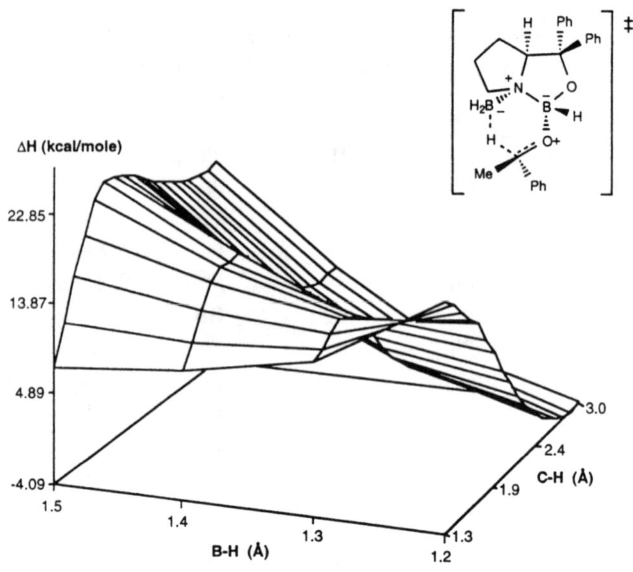

Figure 7. Reaction surface generated for the reduction of acetophenone.

III. RESULTS AND DISCUSSION

These calculations resulted in a three-dimensional plot of heat of formation (ΔH_f^0 kcal/mole) versus B–H distance (Å) versus C–H distance (Å) from which the saddle point corresponding to the transition state was located. (All the surfaces generated were of a very similar nature and have the same general shape. See Figure 7 for the reaction surface obtained for **1**.)

In each case, transition state location was verified by the presence of one negative force constant and zero or very small trivial vibrations resulting from the FORCE calculation. The major component of the negative vibration was found to be due to B–H bond vibration, that is, associated with bond breaking. This confirmed that the correct transition state had been located.

IV. ANALYSIS OF RESULTS

A method was needed to analyze the results which would take into account the relative contribution of product formation via all the possible transition states calculated. For minima population, the Boltzmann distribution can be applied, but when dealing with maxima, Transition State Theory must be applied to this problem.

The general approach to this analysis is based on the following theory. Consider a starting material, A, going via transition state $[A]^{\ddagger}$, to form products with an activation energy of V_0 (Figure 8). The rate of reaction for this process is given by

$$k = (RT/h)\exp(-V_0/RT)(Q_A^{\ddagger}/Q_A) \tag{1}$$

where Q denotes the partition function.

The partition function is the product of translational, rotational, and vibrational components and is simplified in two ways. The first simplification assumes that translational partition functions cancel. Then, by assuming that the differences in the moments of inertia among the alternative transition states are small, the contributions of rotational partition functions compared to the vibrational partition functions and rotational components may be eliminated, giving the simplified form of the rate equation:

$$k = (RT/h)\exp(-V_0/RT)(Q_{A\text{vib}}^{\ddagger}/Q_{A\text{vib}}) \tag{2}$$

All the variables in this expression are now known quantities or are values that can be calculated. For example, the temperature at which the reactions are run, T, is provided in the experimental details; the activation energy, V_0, is determined from the difference in energy between the transition state and starting complex; partition functions, Q, are deter-

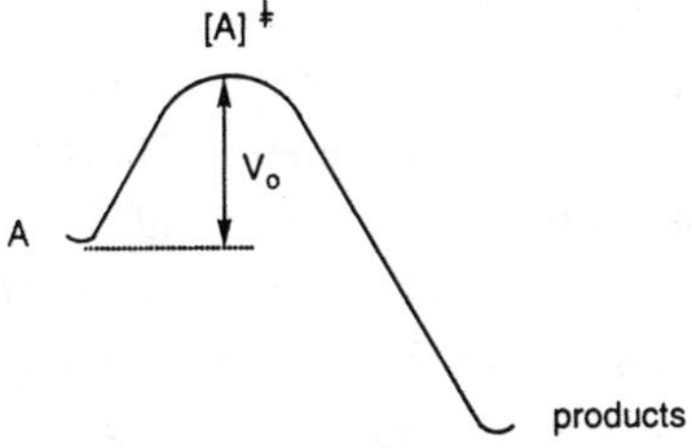

Figure 8. General reaction profile.

mined using the MOPAC keywords THERMO and ROT (which defines the symmetry number of the molecule).

Applying this theory to the case in hand in which two diasteromeric starting complexes each with three distinct transition states (chair, boat, and half-chair), lead to products that became more complex. At this stage in the analysis it was possible to neglect the contribution of the half-chair, since in all cases studied they were more than 5 kcal/mole higher in energy than the lowest energy transition state; thus, its contribution to the final product would be negligible. From Eq. (2), rate equations may be derived for each transition state, which can be substituted into the branching ratio, Eq. (3). Simplification and rearrangement of the branching ratios followed by normalization gave relative values of each rate constant and hence calculated %ee values.

$$\frac{k_n}{k_1 + k_2 + k_3 + k_4} \qquad (3)$$

where

k_1 = rate constant for diasteromeric complex 1 chair
k_2 = rate constant for diasteromeric complex 1 boat
k_3 = rate constant for diasteromeric complex 2 chair
k_3 = rate constant for diasteromeric complex 2 boat

Theoretical ee's calculated using these rate constants agreed quite well with the experimental results (Table 2). In contrast to the boat transition state proposed by Corey et al. and Evans, the transition state with the lowest overall E_a in every case considered was a chair.

To validate the use of MNDO for studying this problem, a comparison was made of the crystal structure geometry of an oxazaborolidine/borane complex[13] with its calculated geometry. The two structures were in excellent agreement with each other; bond lengths varying by only hundredths of an angstrom and bon dangles by less than 5° (see Table 3). Overlaying the heavy atoms of the oxazaborolidine bicyclic ring system in SYBYL gave an RMS deviation of 0.13 Å, showing the similarity between the ring systems for the calculated and crystal structure geometry (see Figure 9).

Table 2. Comparison of Experimental and Calculated Results

	$E_a{}^a$ chair 1	E_a boat 1	E_a chair 2	E_a boat 2	Calc'd[b] %ee	Exper. % ee
1	15.78	16.78	19.32	17.06	98	97.0^c
2	15.58	16.88	17.23	16.74	80	90.0^c
3	15.17	16.51	16.56	17.43	90	92.0^c
4	14.58	16.07	16.63	16.89	93	96.5^c
5[h]	12.48	13.64	14.48	14.46	94	97.6^c
6	13.92	16.09	16.49	17.19	96	97.3^c
7[h]	12.51	15.79	13.88	14.83	88^d	95.6^e
8	17.22	18.14	18.28	19.21	94	97.8^f
9	14.97	21.97	15.19	15.53	59^f	poor
10	17.07	17.55	17.23	17.47	58^f	poor
11	16.92	17.35	16.81	18.13	$11^{f,g}$	poor
12	15.17	17.02	15.11	16.65	$48^{f,g}$	poor
13	16.52	17.06	16.54	17.05	2^f	poor
14	14.21	14.68	14.39	14.79	42^f	poor

Notes: [a]Kcal/mole. [b]Half-chair E_a in all cases more than 5 kcal/mole higher in energy than lowest energy transition state. Its contribution to the final product distribution will be negligible. [c]See references 1a and 1b. [d]Reaction temperature of 253 K. [e]Unpublished results. [f]Assumes a reaction temperature of 263 K. Exact experimental % ee values not given. See reference 1c. [g]The R-enantiomer is the major product in all examples except **11** and **12**. Since it is difficult to systematically search the conformational spalce of transition states which possess several remote rotatable bonds using semi-empirical techniques, the calculated % ee's for these entries should be considered somewhat less reliable.

The major different observed was in the position of the diphenylprolinol rings. In the calculated structure, the phenyl groups were rotated by approximately 15° compared to the crystal structure, positioning these groups farther apart from each other (Figure 9). This could be accounted for by the effect of packing forces in the crystal structure. Corey et al.[14] have published an X-ray crystal structure of the same borane adduct as presented by Mathre.[13] The results from both structures are in good agreement.

Using SYBYL 5.41, the geometry of each transition state was carefully examined, determining atoms and groups that would be most likely to be involved in any steric interactions and also assessing freedom of rotation. This analysis was broken down into the following effects: (1) the orientation of groups R_L and R_S of the ketone; (2) the size of R_S; (3)

Table 3. Comparison of the Key Geometric Features of
the Borane Adduct (see Figure 9) Crystal and Calculated
Structures

	Calculated Structure	*Crystal Structure*
Bond Lengths (Å)		
N_1-B_1	1.54	1.49
N_1-B_2	1.65	1.62
B_1-C_1	1.56	1.54
B_2-H_1	1.18	0.95
B_1-O_1	1.35	1.35
Bond Lengths (°)		
$N_1-B_2-H_1$	107.1	109.5
$B_1-N_1-B_2$	105.1	100.9
$N_1-B_1-O_1$	108.4	110.2
$N_1-B_1-C_1$	129.1	126.6
$B_1-C_1-H_4$	109.5	109.4
Dihedral Angles (°)		
$B_2-N_2-B_1-C_1$	−71.5	−78.7
$B_2-N_2-B_1-O_1$	109.6	108.4

the effect of altering R_L group; (4) the "fine-tuning" effect of R_2, the boron substituent; (5) the importance of the R_1 group.

In the lowest energy transition state for **1** (**chair 1**, Figure 10), the phenyl group occupies the more energetically-favorable *exo* position (i.e., *anti* to R2) because it can lie in a plane parallel to the diphenylprolinol ring at a distance of more than 5.5 Å from the angular phenyl group.[15] Conversely, in the first diastereomeric transition state (**chair 2**, Figure 10), the *exo* methyl group, due to its tetrahedral geometry, is approximately 1.5 Å closer to the phenyl group. Moreover, considerable interactions of the phenyl group with hydrogens on the *endo* (i.e., *syn* to R_2) face of the catalyst also disfavor this arrangement.

Selectivity is considerably more sensitive to the choice of the catalyst than to the nature of the ketone being reduced, although increasing the size of R_S from a methyl to an ethyl group (e.g., **1** and **2**) does decrease selectivity. This is due to the increase in steric bulk on the *endo* face, leading to greater interactions on this face with both $-BH^3$ and $-BH$ moieties and decreased freedom of rotation of R_S. Figure 11 illustrates

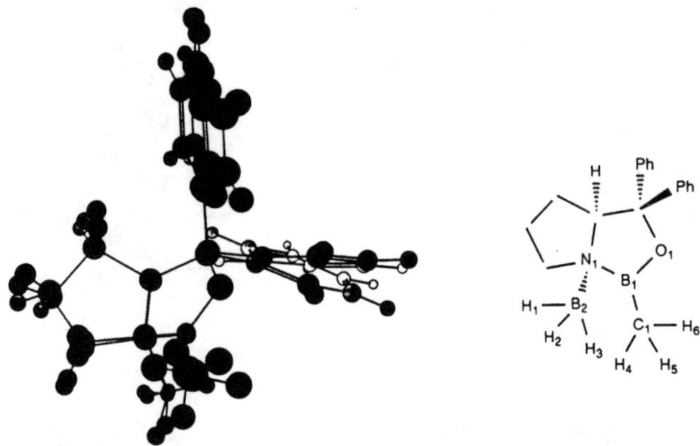

Figure 9. Crystal and calculated structure overlaid.

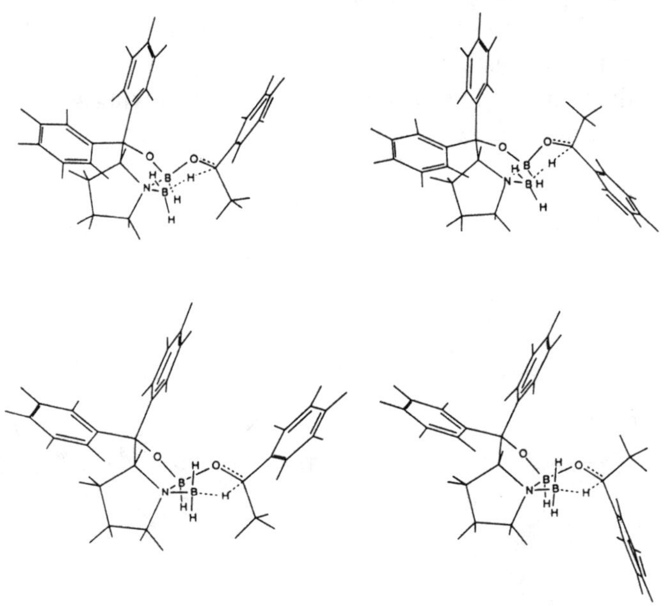

Figure 10. Diastereomeric chair and boat transition states for **1** showing the R_L and R_S group orientation.

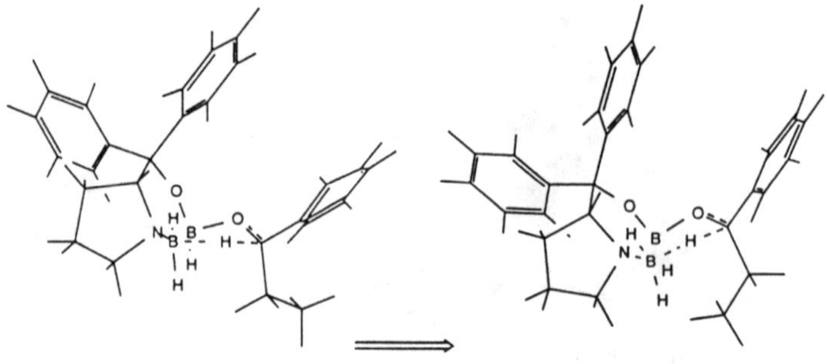

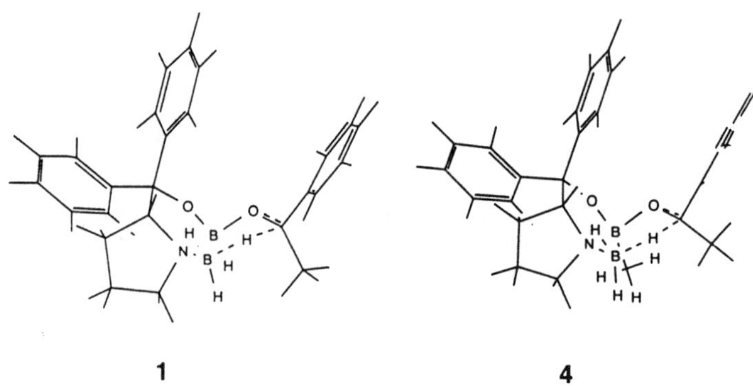

Figure 11. Effect of increasing the R_S group size.

this for the **chair 1** transition state for **2**. This shows that rotation about the C–C bond of the ethyl group from conformation **a** to **b** would lead to increased interactions with the boron substituents.

Variation of the boron substituent (**1** and **4**; **2** and **5**; and **7** and **3**) has a "fine-tuning" effect on the oxazaborolidine performance. Changing from H to Me (see **chair 1** transition states for **1** and **4**, Figure 12) increases the congestion on the *endo* face, amplifying interactions with one of the two ketone substituents and thereby giving greater impetus for R_L to occupy the *exo* face.

Figure 12. "Fine-tuning" effect of boron substituent variation.

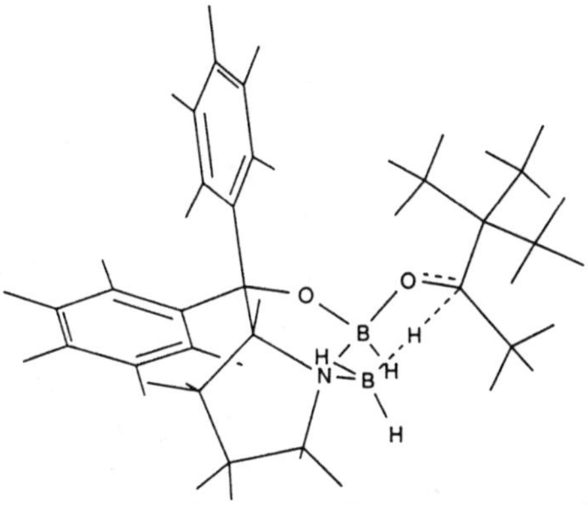

Transition state **chair 1, 3**

Figure 13. Variation in the nature of the R_L group.

The decrease in enantioselectvity observed when changing R_L from an aryl to an alkyl group (e.g., phenyl and t-butyl, **1** and **3**, and **4** and **6**) can be attributed to the branched nature of the alkyl group which results in a decrease of the distance to the diphenylprolinol groups by, on average, 2 Å. This illustrated that, in order to observe high enantioselectivity, the large group (R_L), must be able to exhibit both "small" and "large" properties, that is, large with respect to its positioning preferentially on the *exo* rather than the *endo* face, but small with respect to minimizing its interactions with the diphenylprolinol groups (see Figure 13).

The nature of the prolinol substituents is the single, most important determinant of stereoselectivity. In the effective catalysts these groups are conformationally linked to provide a facial bias which greatly favors one set of diastereomeric transition states over the other. In general, these groups have to be able to maximize the distance to the R_L group and also have freedom of rotation with respect to each other. For example, the phenyl and β-naphthyl groups (**1** to **7**, and **8**) are able to lie parallel to the R_L substituent at a distance of more than 5.5 Å (see Figure 14, **chair 1** transition state for **8**).

Conversely, the less effective groups (**9** to **12**) all lie at approximately 1 Å to 2 Å closer with rotation causing greater interactions with R_L. While these interactions alone do not govern the observed selectivity, they nonetheless appear to parallel the total steric interactions for the various

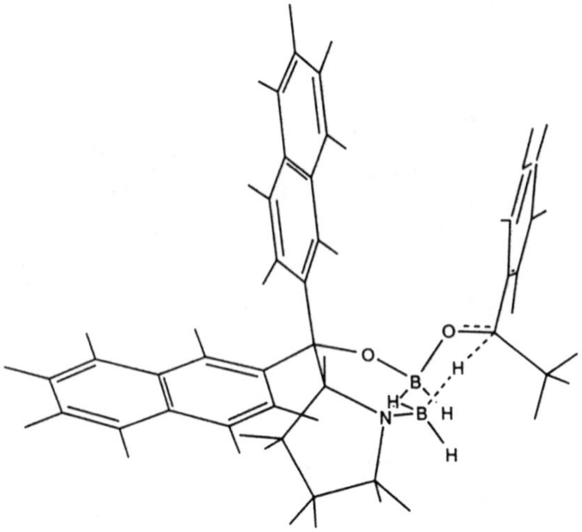

Transition state **chair 1, 8**

Figure 14. β-Napthyl groups as effective substituents.

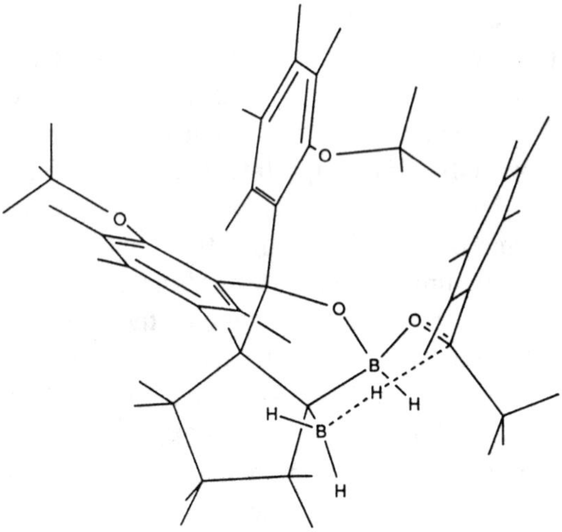

Transition state **chair 1, 11**

Figure 15. *o*-Substituted aryl derivatives as ineffective substituents.

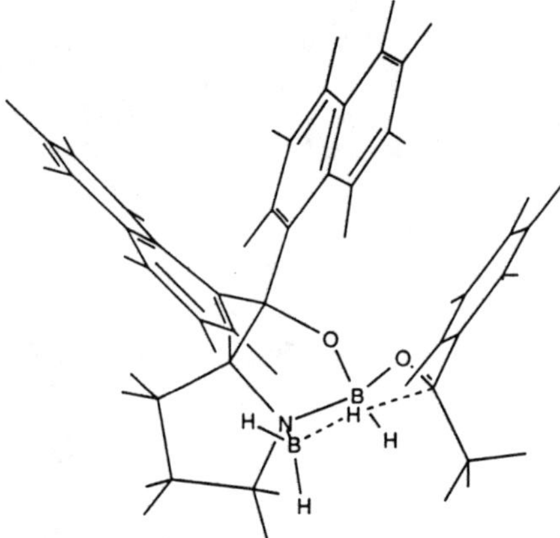

Transition state **chair 1, 10**

Figure 16. α-Napthyl groups as ineffective diprolinol substituents.

transition states. For *o*-substituted aryl derivatives and α-naphthyl groups (e.g., **chair 1** transition states for **10** and **12**, Figures 15 and 16) the groups suffer increased interactions with their geminal counterparts, resulting in the population of rotamers which decreases the difference in

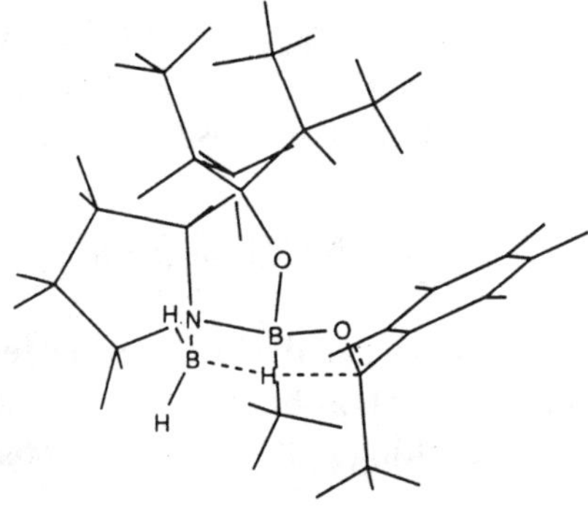

Transition state **chair 1, 9**

Figure 17. Geminal interactions render *i*-propyl an ineffective diprolinol substituent.

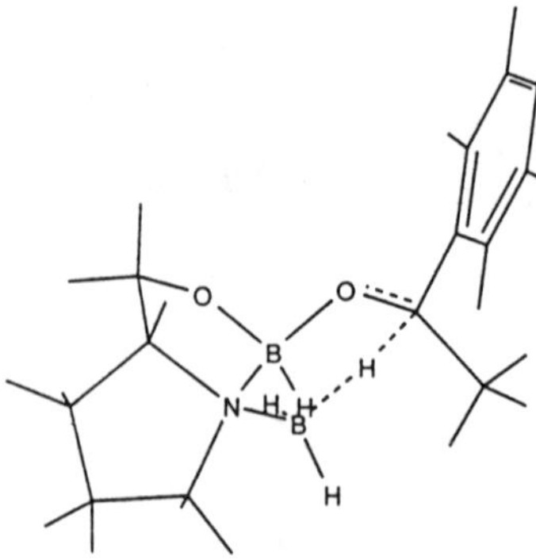

Figure 18. Chair 1 transition state for **13** showing the lack of steric basis associated with this catalyst.

congestion between the *exo* and *endo* faces of the catalyst. In the case of *i*-propyl (e.g., **chair 1** transition state **9**, Figure 17), due to its branched nature, the same detrimental geminal interactions cause this to be an ineffective catalyst.

For the cases when R_1 = H (**13** and **14**), the inherent steric bias associated with the bicyclo [3.3.0] ring system is insufficient on its own to permit any significant facial differentiation (see Figure 18). Thus, the only effective catalysts are those in which the substituents, as a consequence of a conformational synergy, amplify the facial bias which an approaching substrate encounters, thereby increasing the energy differences among the various alternative transition states.

V. CONCLUSIONS

In summary, it has been shown that the hydride transfer occurs via a chair transition state, with the oxazaborolidine and ketone substituents effects reinforcing each other, resulting in the selectivity observed.

REFERENCES AND NOTES

1.   (a) Itsuno, S.; Ito, K.; Hirao, A.; Nakahama, S. *J. Chem. Soc., Chem. Commun.* **1983**, 469; (b) Itsuno, S.; Ito, K.; Hirao, A.; Nakahama, S. *J. Org. Chem.* **1984**, *49*, 555; (c) Itsuno, S.; Nakano, M.; Miyazaki, K.; Masuda, H.; Ito, K. *J. Chem. Soc.*

Perkin Trans. I, **1985**, 2039; (d) Itsuno, S.; Nakano, M.; Ito, K.; Hirao, A.; Owa, M.; Kanda, N.; Nakahama, S. *J. Chem. Soc. Perkin Trans. I*, **1985**, 2615; (e) Itsuno, S.; Sakurai, Y.; Ito, K.; Hirao, A.; Nakahama, S. *Bull. Chem. Soc. Jpn.* **1987**, *60*, 395.

2. (a) Corey, E.J.; Bakshi, R.K.; Shibata, S. *J. Am. Chem. Soc.* **1987**, *109*, 5551; (b) Corey, E.J.; Bakshi, R.K.; Shibata, S.; Chen, C-P.; Singh, V.K. *J. Am. Chem. Soc.* **1987**, *109*, 7925; (c) Corey, E.J.; Gavai, A.V. *Tetrahedron Lett.* **1988**, *29*, 3201; (d) Corey, E.J.; Jardine, P.; Mohri, T. *Tetrahedron Lett.* **1988**, *29*, 6409; (e) Corey, E.J.; Shibata, S.; Bakshi, R.K. *J. Org. Chem.* **1988**, *53*, 2861; (f) Corey, E.J. *Chem. Soc. Rev.* **1988**, *17*, 111; (g) Corey, E.J.; Reichard, G.A. *Tetrahedron Lett.* **1989**, *30*, 5207; (h) Corey, E.J.; Chen, C-P.; Reichard, G.A. *Tetrahedron Lett.* **1989**, *30*, 5547; (i) Corey, E.J.; Link, J.O. *Tetrahedron Lett.* **1989**, *30*, 6275; (j) Corey, E.J.; Link, J.O. *Tetrahedron Lett.* **1990**, *31*, 601; (k) Corey, E.J.; Link, J.O. *J. Am. Chem. Soc.* **1992**, *114*, 1906; (l) Corey, E.J.; Link, J.O. *Tetrahedron Lett.* **1992**, *33*, 3431.

3. (a) Rao, A.V.R.; Gurjar, M.K.; Sharma, P.A.; Kaiwar, V. *Tetrahedron Lett.* **1990**, *31*, 2341; (b) Rao, A.V.R.; Gurjar, M.K.; Kaiwar, V. *Tetrahedron Asymmetry* **1992**, *3*, 859.

4. Jones, T.K.; Mohan, J.J.; Xavier, L.C.; Blacklock, T.J.; Mathre, D.J.; Sohar, P.; Jones, E.T.T.; Remear, R.A.; Roberts, F.E.; Grabowski, E.J.J. *J. Org. Chem.* **1991**, *56*, 763.

5. Tanaka, K.; Matsui, J.; Suzuki, H. *J. Chem. Soc., Chem. Commun.* **1991**, 1311.

6. Kim, Y.H.; Park, D.H.; Byun, I.S.; Yoon, I.K.; Park, C.S. *J. Org. Chem.* **1993**, *58*, 4511.

7. (a) Nevalainen, V. *Tetrahedron Asymmetry* **1991**, *2*, 63; (b) *Tetrahedron Asymmetry* **1991**, *2*, 429; (c) *Tetrahedron Asymmetry* **1991**, *2*, 827; (d) *Tetrahedron Asymmetry* **1991**, *2*, 1133; (e) *Tetrahedron Asymmetry* **1992**, *3*, 921; (f) *Tetrahedron Asymmetry* **1992**, *3*, 933.

8. Evans, D.A. *Science*, **1988**, *240*, 420.

9. (a) QCPE Program No. 455, Version 5.0; (b) Jones, D.K.; Liotta, D.C.; Shinkai, I.; Mathre, D.J. *J. Org. Chem.* **1994**, *116*, 8516; (c) Quallich, G.J.; Blake, J.F.; Woodall, T.M. *J. Am. Chem. Soc.* **1994**, *116*, 8516; (d) Linney, L.P.; Self, C.R.; Williams, I.H. *J. Chem. Soc., Chem. Commun.* **1994**, 1651.

10. We are grateful to Serena Software for making this software available to us.

11. Gilbert, K.; Gajewski, R. *Advances in Molecular Modeling*, Vol. 2, D. Liotta (ed.), JAI Press, Greenwich, CT, 1990.

12. (a) Dewar, M.J.S.; Thiel, W. *J. Am. Chem. Soc.* **1977**, *99*, 4899. (b) Dewar, M.J.S.; McKee, M.L. *J. Am. Chem. Soc.* **1977**, *99*, 5231.

13. Mathre, D. 203rd American Chemical Society National Meeting, San Francisco, Abstract 85.

14. Corey, E.J.; Azimioara, M.; Sarshar, S. *Tetrahedron Lett.* **1992**, *33*, 3429.

15. All distances are measured from center to center for aryl groups and the closest proton is the point of measurement for alkyl groups.

SEMIEMPIRICAL MOLECULAR ORBITAL METHODS

D. Quentin McDonald

Advances in Molecular Modeling
Volume 3, pages 21–46.
Copyright © 1995 by JAI Press Inc.
All rights of reproduction in any form reserved.
ISBN: 1-55938-326-7

I. INTRODUCTION

Semiempirical methods are those which combine simplified quantum theory and experimentally derived parameters with the aim of allowing the computation of molecular properties (energies, geometries, electronic structure, etc.) to a high degree of accuracy and with minimal computational effort. This review is a discussion of the development of some widely used semiempirical molecular orbital methods, with particular emphasis on those which contributed to the development of the modern methods.

II. WHERE IT ALL BEGAN

Semiempirical methods have a foundation in common with all quantum chemistry, namely the work of Erwin Schrödinger, which was carried out in the early part of this century. He built on the wave mechanics of de-Broglie and the atomic structure proposed by Bohr to develop the quantum mechanical description of a set of interacting particles, that is, the nuclei and electrons in a molecule. This is succinctly represented by the time independent Schrödinger wave equation (SWE):

$$\{T + V\}\Psi = E\Psi \tag{1}$$

The SWE relates the linear operators for kinetic and potential energy, T and V, respectively; Ψ is the "wave function" which describes the spatial distribution of the particles; and E, the total energy of the system.

A direct solution of the SWE is possible only for one-electron systems such as the hydrogen atom. In order to apply this theory to multi-electron systems a series of approximate theoretical treatments are applied. The basis of these is the Hartree–Fock approximation. Here the total energy of a system can be described as the sum of a number of terms:

$$E = \sum_i^n h_i + \frac{1}{2} \sum_{ij}^n (J_{ij} - K_{ij}) + \sum_{\kappa\lambda}^n Z_\kappa Z_\lambda / (R_\kappa - R_\lambda) \tag{2}$$

The h_i term represents the kinetic energy of the electron, the interaction of the electron with the nucleus, and is given by:

$$h_i = \int \phi_i(k)\left(-\tfrac{1}{2}\nabla_k^2 - \sum \frac{Z_k}{(r-R_k)}\right)\phi_i(k)d\tau \tag{3}$$

The *J* term is a classical representation of the electrostatic charge repulsion between the two charge distributions and is given by:

$$J_{ij} = \int\int \phi_i^*(k)\phi_i(k)(R_k-R_l)^{-1}\phi_j^*(l)\phi_j(l)d\tau d\tau \tag{4}$$

This classical repulsion is modified by the K_{ij} term which represents the "exchange integral":

$$K_{ij} = \int\int \phi_i^*(k)\phi_j(k)(R_k-R_l)^{-1}\phi_j^*(l)\phi_i(l)d\tau d\tau \tag{5}$$

which represents the bonding contribution between electron distributions.

Values for ϕ_i are determined by minimizing the total energy with respect to variations in ϕ_i. By the "variation principle" this gives the best possible set of orbitals ϕ, which is known as the Hartree–Fock wavefunction. The minimization of the energy can be expressed in terms of n simultaneous equations of the form:

$$F(k)\phi_i(k) = \varepsilon_i\phi_i(k) \tag{6}$$

Here $F(k)$ is the Fock operator and is the sum of a one-electron term, $H(k)$, and a multi-electron term $G(k)$. Generally the solution of ϕ_i is determined by assuming some form of the starting orbitals, substituting the appropriate calculated values into Eq. (6) to obtain a further set and repeating in an iterative process until no change is observed in the orbitals; the system has then become "self-consistent," and so this method is known as the Self Consistent Field (SCF) method. Thus the detailed treatment of the repulsions of every pair of electrons required in the SWE has been approximated by a treatment that considers the average field that each electron exerts on all other electrons. This process, however, makes no allowance for the fact that in multi-electron systems electron motion is not independent. The motion of the electrons is correlated in order to ensure that the inter-electron repulsion is kept to a minimum. The SCF procedure, therefore, cannot be used to calculate accurately the total energy of a system. The value obtained will always be in error by an amount that is, by definition, the correlation energy of the system. Semiempirical methods compensate for electron correlation by the inclusion of parameters in the expressions for H and G which are derived from experimentally derived quantities.

The Hartree–Fock equation can be solved numerically for atoms, but in molecular systems with lower symmetry this is not practical. A milestone in quantum chemistry was reached in the early 1950s when Roothaan[1] and Hall[2] each independently developed an implementation of the variation method which involves the use of the atomic orbitals, ϕ, in a molecule as the basis functions Φ. This treatment is known as the Linear Combination of Atomic Orbitals–Self Consistent Field (LCAO–SCF) method. In a molecule the electrons are considered to occupy, in pairs, a set of molecular orbitals ψ_j. First the electrons μ and ν are considered to reside in atomic orbitals ϕ_μ and ϕ_ν centered on atom A, and λ and σ reside in ϕ_λ and ϕ_σ on atom B. So, with this notation for n atomic orbitals,

$$\psi_i = \sum_i^n \phi_\nu C_{\nu i} \tag{7}$$

and:

$$\sum C_i (F_{\mu\nu} - S_{\mu\nu}\varepsilon_i) = 0 \tag{8}$$

The term $S_{\mu\nu}$ is known as the "overlap integral" and is defined as:

$$S_{\mu\nu} = \int \phi_\nu^* \phi_\mu \, d\tau \tag{9}$$

The multi-electron Fock matrix $F_{\mu\nu}$ can be considered the sum of a one-electron term and a two-electron term as discussed above. So,

$$F_{\mu\nu} = H_{\mu\nu} + G_{\mu\nu} \tag{10}$$

where $H_{\mu\nu}$ is the matrix element of the one-electron Hamiltonian which includes the kinetic and potential energy for electrons in the field of the core. So:

$$H_{\mu\nu} = \int \phi_\mu \left[-\tfrac{1}{2}\nabla^2 - \Sigma V_a(R) \right] \phi_\nu \, d\tau \tag{11}$$

$G_{\mu\nu}$ is the matrix element which depends on all valence electrons and so is written:

$$G_{\mu\nu} = \sum_{\lambda\sigma} P_{\lambda\sigma} \left[(\mu\nu/\lambda\sigma) - \tfrac{1}{2}(\mu\sigma/\mu\lambda) \right] \qquad (12)$$

$G_{\mu\nu}$ depends on the molecular orbitals as indicated by the presence of the density matrix $P_{\lambda\sigma}$ in Eq. (12).

$$P_{\lambda\sigma} = 2 \sum_i C_{i\lambda} C_{i\sigma} \qquad (13)$$

The two-electron integrals are defined as:

$$(\mu\nu/\lambda\sigma) = \int\int \phi_\nu(1)\phi_\mu(1) \frac{1}{R_{12}} \phi_\lambda(2)\phi_\sigma(2)\, d\tau \qquad (14)$$

There are a large number of such two-electron multidimensional integrals to evaluate in calculations of polyatomic species. Consequently, these are prime targets for approximation in semiempirical treatments.

Finally, the electronic energy of a system is given by,

$$E_{el} = \frac{1}{2} \sum_{\mu\nu} P_{\mu\nu}(F_{\mu\nu} + H_{\mu\nu}) \qquad (15)$$

and the total energy is the sum of the electronic energy and the total repulsion energy between the cores.

$$E_{tot} = E_{el} + \sum_A^N \sum_B^N Z_A Z_B / R_{AB} \qquad (16)$$

With the development of this Roothaan–Hall formalism it was now possible to begin direct meaningful calculations of the wavefunction (and consequently the properties) of a molecular system. Soon after the work of Roothaan and Hall, Robert Parr[3] described a "fundamental dichotomy." The elements of the Fock matrix in Eq. (12) can be calculated from a knowledge of the basis functions—the atomic orbitals. This theoretical approach became known as the ab initio (meaning quite literally "from the start") approach, the calculation of molecular properties from the fundamental constants of physics. It would seem reasonable that this approach should be the aim of everyone attempting to use calculations for chemical applications. There are, however, two important problems. The first is that ab initio calculations are still dependent on the LCAO–SCF method and so their accuracy is bound by the limits of this approximation. Consequently, the results of ab initio calculations are often not sufficiently accurate to be chemically useful, particularly if a minimal basis set of atomic orbitals is used. More importantly, the time taken for ab initio calculations is considerable as a large number of electron interaction integrals must be evaluated. For this reason the calculations before the development of high speed computers were limited to simple diatomic molecules.

There is a demand from chemists to be able to calculate the properties of a wide range of molecules, with reasonable accuracy and at modest cost. For this reason methods were developed which use less rigorous approximations than the Roothaan–Hall treatment, but include the use of experimentally derived parameters to allow calculations to proceed economically and with useful accuracy. These are the semiempirical molecular orbital methods and are the topic of the remaining discussion.

III. THE HÜCKEL MOLECULAR ORBITAL THEORY (HMO)

The Hückel molecular orbital theory[4] is the oldest semiempirical molecular theory, dating back to the 1930s. It is suitable in its most basic form only for large, planar, conjugated molecules. Central to the Hückel theory is the concept that in these molecules the interaction between the π electrons and the σ electrons is small. If this is so then the π electrons may be treated as a loosely bound group of electrons moving in the field of a core made up of the σ electrons and atomic nuclei. This is a one-electron treatment with the multi-electron matrix element F_{ij} in Eq. (11) being approximated by a one-electron term H_{ij}^{π}. In its most basic form the HMO theory applies four approximations during the evaluation of the one-electron matrix. The first is that of zero differential overlap. That is to say, the overlap integral S_{ij} as defined in Eq. (17) is treated as,

$$S_{ij} = \delta_{ij} \tag{17}$$

where δ_{ij} is known as the Kronecker delta and $\delta_{ij} = 1$ if $i = j$ and $\delta_{ij} = 0$ if $i \neq j$. Second, the diagonal terms H_{ii}^{π} which represent the coulombic integrals in the molecule are set to a common value for every equivalent center in the molecule such that:

$$H_{ii}^{\pi} = (\phi_i / H^{\pi} / \phi_j) = \alpha \tag{18}$$

The resonance integral is also set to a standard value when the atomic orbitals are on neighboring atoms:

$$H_{ij}^{\pi} = (\phi_i / H^{\pi} / \phi_j) = \beta \qquad \text{if } i,j \text{ are adjacent atoms} \tag{19}$$

The fourth and final approximation is that the off diagonal terms are set to zero if the atomic orbitals in the basis set are not located on neighboring atoms.

$$H_{ij}^{\pi} = 0 \qquad \text{if } i, j \text{ are not adjacent atoms} \tag{20}$$

Although it may appear that the quantities α and β may be evaluated explicitly from the above expressions, this is not the case as the Hamiltonian H^{π} is not well defined. Instead, the secular equation which results with these quantities is solved to give values of the roots (the orbital energies) in terms of the parameters α and β. These parameters can then be adjusted to fit experimental orbital energies obtained from spectroscopic or polarographic experiments. Once suitable values of α and β are determined, the coefficients C_{ij} can be established. This results in an approximate description of the wavefunction for the π system. There is, however, very little in the HMO theory that can be described as theoretically sound. The neglect of π/σ interaction, antisymmetrization, and overlap influences means that it is really only useful for correlative purposes. Furthermore, the Hückel approach is best in the situation where there are a large number of compounds of a similar type to allow accurate parameterization. For this reason heteroaromatic compounds prove a particularly difficult problem; not only do they require a larger number of parameters, but there are smaller numbers of heteroaromatic compounds available to provide the necessary experimental data for parameterization. Notwithstanding the deficiencies of the HMO method, it paralleled the picture of molecular orbitals held by organic chemists of the time and so it was highly utilized until the early 1960s. At that time

Hoffmann[5] commented that, "The steady pursuit of correlations between theoretically calculated π electron properties and measurables has unfortunately cast a shadow of unreality on the σ framework." A need for a more general semiempirical treatment was evident.

IV. EXTENDED HÜCKEL MOLECULAR ORBITAL THEORY

The logical progression from the simple HMO theory was to attempt to increase its generality of application while maintaining the ease of calculation. The extended Hückel molecular orbital theory[6] was based on earlier work by Mulliken, Wolfsberg, and Helmholtz to allow calculations for aromatic and aliphatic, organic and inorganic systems, with one simple parameterization. In the extended Hückel molecular orbital theory the zero differential overlap approximation is no longer applied. Rather, the overlap integral is evaluated exactly using Slater type orbitals for a valence only basis set. The critical difference is, however, that although it is still essentially a one-electron approximation to the full Fock matrix, the matrix elements representing coulombic repulsion, H_{ii}, are evaluated directly from experimental values for the valence state ionization potentials (VOIP) so that $H_{ii}(\text{H1s}) = -13.6\text{eV}$, $H_{ii}(\text{C2s}) = -21.4\text{eV}$ and $H_{ii}(\text{C2p}) = -11.4\text{eV}$. The off-diagonal terms (the resonance integrals) are approximated by a function of the coulombic terms and the overlap integral:

$$H_{ij} = 0.5K(H_{ii} + H_{jj})S_{ij} \qquad \text{where } K = 1.75 \qquad (21)$$

This treatment was therefore one of the first to apply systematic parameters in an early stage of the calculation. With computer implementation the Hückel energy of a molecule could be evaluated quickly, allowing the study of the variation of potential energy with geometry. An example is the C–H bond of methane where the potential energy curve has the correct form but the minimum in the potential well occurs at about 1.0Å, which is too short for such a bond. This highlights the problems of such a one-electron treatment in that it could not correctly predict bond lengths, binding energies, or molecular orbital energies. If idealized bond lengths were assumed, then the extended HMO treatment could be successful in predicting the conformational preferences for cyclohexane and obtaining a value close to the experimental value for the barrier to rotation in ethane.

Subsequent work on the extended HMO theory has attempted to refine the parameters by adjustment to mimic ab initio calculations[6] which, however, was not been able to extend the accuracy of such procedures to match the ab initio calculations, and it was clear that a different approach to simplification of the secular Eq. (17) was required.

V. COMPLETE NEGLECT OF DIFFERENTIAL OVERLAP (CNDO)

In 1965 Pople and co-workers, recognizing the limitations of the Hückel based approaches, developed the Complete Neglect of Differential Overlap (CNDO) scheme.[7] They developed a sophisticated approach to introduce parameters into quantum theory which has formed the basis for today's modern semiempirical molecular orbital methods. The Pople CNDO method of the mid-1960s was based on the earlier π-only treatment of Pariser, Pople and Parr[8] which did allow for some two-electron interaction. This earlier theory was the first "CNDO" treatment as it neglected all two-electron integrals which depend on overlap of charge in different orbitals, the remaining integrals being calculated by the use of semiempirical expressions. This approach was successful in prediction of electronic spectra,[9] and the calculation of some molecular properties, for planar unsaturated molecules. While it was definitely an improvement on simple Hückel theory, it was still not generally applicable to problems of wider chemical interest.

Pople recognized that any approximation with the intention of simplifying the quantities in Eq. (8) must retain invariance with respect to transformation of the atomic orbitals used in the basis sets. The results of a calculation must be the same after changing the local axes: that is, a rotational invariance. They must also be invariant to the transformations involving mixing of s and p orbitals. With these conditions in mind he proposed the CNDO method for approximate calculation of the quantities listed in Eqs. (9)–(14). There are essentially five approximations central to the CNDO theory:

1.   Complete neglect of differential overlap. The atomic orbitals are treated, as in the Hückel treatment, as an orthonormal set with $S_{\mu\nu} = \delta_{\mu\nu}$.

The result of this treatment is that the coefficients form an orthogonal matrix:

$$\sum_{\mu} C_{\mu i} C_{\mu j} = \delta_{ij} \tag{22}$$

For different orbitals on the same atom the overlap is already zero, but for different orbitals on different atomic centers this is a severe approximation.

2.   All the two-electron integrals which depend on overlap of charge densities of different basis orbitals are neglected so:

$$(\mu\nu/\lambda\sigma) = 0 \quad \text{unless } \mu = \nu \text{ and } \lambda = \sigma \tag{23}$$

The non-zero values are given a common value:

$$\gamma_{\lambda\mu} = (\lambda\lambda/\mu\mu) \tag{24}$$

This approximation does not satisfy the requirement for rotational invariance and a third approximation is therefore required.

3. $\gamma_{\lambda\mu}$ is assumed to depend only on the type of atoms to which ϕ_{μ} and ϕ_{λ} belong and not on the type of orbital. The quantity γ_{AB} measures the average electron repulsion between an electron on atom A and an electron on atom B and this averaging of the repulsion restores rotational invariance. γ_{AB} is calculated from the use of Slater S atomic orbitals for atoms A and B in the following expression:

$$\gamma_{AB} = \int \int S_A^2(1) \frac{1}{R_{AB}} S_B^2(2) d\tau \tag{25}$$

With these approximations the expression for the diagonal elements of the Fock matrix becomes,

$$F_{\mu\mu} = H_{\mu\mu} - P_{\mu\mu}\gamma_{AA}P_{AA}\gamma_{AB} + \sum_{A \neq B} P_{BB}\gamma_{AB} \tag{26}$$

where P_{BB} is the total valence electron density on B and $H_{\mu\mu}$ is defined as,

$$H_{\mu\mu} = \left(\mu/-\tfrac{1}{2}\nabla^2 - V_A/\mu\right) - (\mu/V_B/\mu)$$

$$= U_{\mu\mu} - \sum(\mu/V_B/\mu) \tag{27}$$

where $U_{\mu\mu}$ is an atomic parameter and can be derived from experimentally measurable values for atomic energy levels.

4.   The quantity $(\mu/V_B/\mu)$ represents the interaction of ϕ_μ with the core of all other atoms. In the CNDO treatment it is set to a value which is the same for all the valence atomic orbitals. So,

$$(\mu/V_B/\mu) = V_{AB} \tag{28}$$

and V_{AB} is calculated directly from the integral,

$$V_{AB} = \int S_{A^2}(1) Z_B/R_{AB} d\tau \tag{29}$$

based on a classical point charge treatment (S_A is the Slater S-type atomic orbital on atom A), and so:

$$H_{\mu\mu} = U_{\mu\mu} - \sum_{B(\neq A)} V_{AB} \tag{30}$$

5.   The off-diagonal one-electron terms are the subject of the fifth and last approximation. If μ and ν are on the same atom, then $H_{\mu\nu}$ is set to zero. If they are on different atoms then $H_{\mu\nu}$ represents the interaction of the charge distribution $\phi_\mu\,\phi_\nu$ with adjacent cores, a bonding contribution is known as the "resonance integral." In the CNDO treatment resonance integral is considered to depend only on the local environment, so,

$$H_{\mu\nu} = \beta_{\mu\nu} = \beta_{AB\circ} S_{\mu\nu} \tag{31}$$

where $\beta_{AB\circ}$ is a parameter for each bond type. This parameter was determined by a trial and error method to give the best fit to the results of full LCOA–SCF calculations.

Calculations using the CNDO method were carried out[10] on a range of diatomic and small polyatomic molecules. The results were disappointing, leading to poor bond lengths and dissociation energies. These failings suggest that the errors introduced in the neglect of differential overlap are a function of internuclear distance. The calculated equilibrium bond angles are more satisfactory where there is less dependence on the neglect of differential overlap. The CNDO method is able to predict the correct stereochemistries for methanol and ethanol. The main success of the CNDO method has been that it served as a starting point for the development of a useful semiempirical method. It demonstrated that such methods require less computational effort than a full LCAO–SCF calculation, but can provide useful structural information.

A small modification to the CNDO scheme, known as CNDO/2,[11] was introduced by Pople the following year. This modification involves the neglect of the so-called penetration integrals. In the original CNDO the diagonal terms in the Fock matrix could be written:

$$F_{\mu\mu} = U_{\mu\mu} + \left(P_{AA} + \tfrac{1}{2}P_{\mu\mu}\right)\gamma_{AA} + \sum_{B(\neq A)} (P_{BB} - Z_B)\,\gamma_{AB}$$

$$+ \sum_{B(\neq A)} (Z_B\gamma_{AB} - V_{AB}) \tag{32}$$

The final term in Eq. (32) represents the effect of electrons in one orbital penetrating those of another orbital leading to a nett attraction. As CNDO/1 binding energies are too large and bond lengths too short, an attempt was made to correct this in CNDO/2 by setting V_{AB} equal to $Z_B\gamma_{AB}$ and thus eliminating the term representing the penetration effects. A further correction involved a superior method of estimating $U_{\mu\mu}$ from atomic data such that,

$$U_{\mu\mu} + \left(Z_A - \tfrac{1}{2}\right)\gamma_{AA} = \tfrac{1}{2}\left(I_\mu + A_\mu\right) \tag{33}$$

where I_μ is the atomic ionization potential and A_μ the electron affinity for the atomic orbital ϕ_μ.

The CNDO/2 method was a slight improvement. Long-range interactions and angular dependence were the only molecular properties being properly treated. There are serious deficiencies in the CNDO method. For example, there is no separation in open shell systems of states arising from the same configuration: methylene therefore shows no separate singlet or triplet states. Furthermore, in many situations and, in particular, planar aromatic molecules, no spin density is calculated in the σ orbitals.

This is the result of the complete neglect of one-center exchange integrals. Pople[12] and Dixon[13] independently recognized this and proposed an improved method in 1967. Their treatments are essentially identical and, for the sake of continuity, only the Pople scheme is discussed in detail here.

VI. INTERMEDIATE NEGLECT OF DIFFERENTIAL OVERLAP (INDO)

In this method a limited number of one-center exchange products are evaluated using a semiempirical expression. The assumption is made that the 2s and 2p orbitals have the same radial parts and the expressions of Slater are used to evaluate the integrals:

$$(ss/ss) = F^0 = \gamma_{AA} \tag{34}$$

$$(sx/sx) = 1/3G^1 \tag{35}$$

$$(xx/xx) = 3/25F^2 \tag{36}$$

Similar expressions are used to evaluate (yy/yy) and so forth. The Slater parameters F^0, G^1, and F^1 were derived from experimental atomic data. This method was originally applied to a range of AB and AB_2 molecules and was successful in correcting the worst deficiencies of the CNDO method which were due to the neglect of the one-center exchange integrals. For example, the correct triplet–singlet separation for methylene is calculated by the INDO method. INDO also showed improvement in the treatment of spin densities, and it was possible to calculate hyperfine coupling constants for ethyl and methyl radicals which were in good agreement with experiment.

The greatest limitation, however, of the INDO approach was that, while some atomic experimentally derived parameters were used, other adjustable parameters were fitted so as to reproduce the results of ab initio calculations. There are two problems with this approach. First, the ab initio calculations were limited at the time to diatomic and small polyatomic molecules and so the information on a large group of molecules needed for complete parameterization was not available. Second, ab initio calculations are themselves limited by the accuracy of the RH method and any treatment attempting to mimic these results will therefore be limited.

A decided change in the philosophy of semiempirical molecular orbital methods occurred in 1968 when Dewar took the INDO treatment and introduced a system where parameters were introduced with the sole purpose of reproducing experimental results.

VII. THE MINDO APPROACH ('MODIFIED INDO')

Between 1968 and 1975 the MINDO approach was developed by the research group led by M.J.S. Dewar. Here, as noted earlier, the primary objective was to calculate with chemical accuracy (± 5.0 kJ mol^{-1}) the ground state equilibrium heat of formation at 25 °C. It was also the aim to be able to determine other useful molecular properties such as geometry, dipole moments, and ionization potentials. Dewar envisaged a "computational spectrometer"; he wished to make computational chemistry as much a part of the everyday work of chemists as NMR or infrared spectroscopy.

The original MINDO approach[14] retained the essence of the Pople INDO method but introduced more parameters in order to try to reproduce experimental results. The major differences were the evaluation of the two-center electron repulsion and the resonance integrals. As in the INDO method the two-center integrals between atomic orbitals on a given pair of atoms is assigned the value γ_{AB}. But instead of evaluating these integrals from Eq. (25) using the Slater type s-orbital functions, a parameterized function was used. The function had to meet two conditions. At infinite separation it goes to the classical e^2/R_{AB} and it should tend to the average one-center value, $\frac{1}{2}(F_A^0 + F_B^0)$ as R_{AB} tends to zero. The expression chosen was,

$$\gamma_{AB} = -14.399 \left[\left((R_{AB})^2 + \rho_A + \rho_B \right)^2 \right]^{1/2} \tag{37}$$

where $\rho_x = -7.1995/F_x^0$

The advantage of evaluating the resonance integral by a semiempirical expression involving experimentally derived parameters is that the effect of electron correlation will implicitly be allowed for in the parameterization. The integrals calculated in this manner are consistently smaller than the corresponding values calculated by the direct evaluation of Eq. (25).

The resonance integrals are also evaluated from a semiempirical function:

$$\beta_{\mu\nu} = S_{\mu\nu}\left(I_\mu^A + I_\nu^B\right)f(R_{AB}) \tag{38}$$

where I_μ is the ionization potential for removal of an electron from the appropriate atomic orbital, $S_{\mu\nu}$ is the overlap integral, and $f(R_{AB})$ is a parameterized function which displays the correct dependence on the inter-center distance R_{AB}. It has the form,

$$f(R_{AB}) = \beta_{cc}^I + (\beta_{cc}^{II}/R_{AB}) \tag{39}$$

for carbon–carbon bonds and,

$$f(R_{AB}) = \beta_{xx}^I \tag{40}$$

for other bonds, where β^I and β^{II} are adjustable parameters.

Finally, the total core–core repulsion was set equal to the total electron–electron repulsion, implying that the total coulombic interaction between two neutral atoms is zero.

MINDO was implemented in a computer program and a number of parameterization schemes investigated. Because of the setting of the total core–core repulsion to the total electron–electron repulsion, it was not possible to determine a single set of parameters which would correctly reproduce experimental bond lengths and heats of formations. Consequently, all MINDO results were for systems with the geometries fixed at their experimental values. Calculations for a range of hydrocarbons and compounds containing nitrogen and oxygen resulted in a RMS error in the calculated heat of formations of around 4.8 kJ mol^{-1}. The necessity of using fixed experimental geometries severely limited the usefulness of the method. This was not the case in the next development in the MINDO series.

MINDO/2[15] evaluates the one- and two-center integrals in the same manner as in MINDO/1, but in this more advanced method the total core–electron attractions are set to equal the negative of the total electron–electron repulsions. The resonance integrals and the core–core repulsions are evaluated by semiempirical expressions that depend on the internuclear separation for the resonance integrals:

$$\beta_{\mu\nu} = BS_{\mu\nu}(I_\mu + I_\nu)f_1(R_{AB}) \tag{41}$$

where I is the atomic ionization potential and B is an adjustable parameter. The core-repulsion between two atoms CR_{AB} was evaluated as,

$$CR_{AB} = ER_{AB} + (Z_A Z_B e^2/R_{AB} - ER_{AB})f_2(R_{AB}) \tag{42}$$

where ER_{AB} is the total electron–electron repulsion between atoms A and B and Z_x is the core charge of X (defined as the difference between the atomic number and the number of valence electrons). The function $f_2(R_{AB})$ must tend to zero at infinite separation (i.e., no bonding) and to unity as the separation approaches zero.

In order to find suitable functions for $f_1(R_{AB})$ and $f_2(R_{AB})$ over 150 combinations were tried. The best results were obtained with $f_1 = 1$ and $f_2 = \exp(-\alpha R_{AB})$. Thus, for every atom pair there are two parameters α and β. Systematic optimization of these parameters resulted in a method which could produce the best agreement with experiment of any of the previous semiempirical methods, although "chemical accuracy" was not achieved.

A significant improvement occurred with the introduction of the final MINDO scheme, MINDO/3,[16,17] which is still in use today.[18] In MINDO/3 a different method is utilized for the evaluation of the one-center integrals $g_{\mu\nu}$ and $h_{\mu\nu}$. Instead of using the Slater–Condon parameters, a method is used which allows independent evaluation of these quantities from atomic data. MINDO/3 also uses a slightly different method for the core–core repulsion function in the case of N–H and O–H bonds.

$$f_2(R_{AB}) = \alpha_{HX}e^{-R_{AB}} \tag{43}$$

The method of parameterization and its implementation has meant that MINDO/3 performs better than any other MINDO scheme. The choice of parameters was carried out almost automatically using a computer program and an iterative least-squares method. MINDO/3 was the first semiempirical method to take full advantage of the techniques for determination of equilibrium geometries by the optimization of geometrical variables in order to minimize the energy of the molecule. The most efficient methods for doing this require both the evaluation of the energy for each geometry and the derivative of the energy with respect to geometrical variables. This latter quantity is easily calculated in the MINDO approximation and thus full geometry optimizations of quite large molecules, for example LSD[19] (144 independent variables) can be carried out without any assumptions being made about the geometry. The MINDO/3 method gives satisfactory results for heats of formations. The geometries and dipole moments calculated by MINDO/3 are in reasonable agreement with experiment. The latter results are particularly interesting as dipole moments were not explicitly included in the parameterization. MINDO/3 was therefore the first semiempirical method to show genuine utility for the calculation of ground state molecular properties. Furthermore, because many possible geometries could be calculated quickly, this method soon began to be used for the study of transient species in chemical reactions, where the study of potential surfaces for a molecule or reacting system may require thousands of calculations.

There are, however, some known problems with the MINDO/3 method. Molecules containing triple bonds are predicted to be too stable, while strain energy is often underestimated in calculations involving the MINDO/3 method. Results for molecules such as H_2N-NH_2, where there are unshared lone pairs on adjacent atoms, are often poor. It is apparent that any method based on the INDO approximation, with the neglect of one-center differential overlap, will not be able to achieve the high degree of accuracy required by experimental chemists.

An alternative, more rigorous, scheme was proposed by Pople in 1968, at the same time as INDO. This was based on the neglect of diatomic differential overlap (NDDO).[10] In the INDO approximation the integrals $(\mu\mu/\sigma\sigma)$ are set equal to some common value γ_{AB} regardless of whether ϕ_μ and ϕ_ν are of the s, $p\sigma$ or $p\pi$ type. In fact, they are not equal and in the NDDO approximation additional bicentric integrals are evaluated which are neglected in the INDO and CNDO schemes. For a pair of dissimilar first row elements there are in fact 22 bicentric integrals to be calculated, while for the simpler treatments there would be only one. Early implementations of the NDDO approximation[20] attempted to calculate the additional bicentric orbitals from direct quadrature of Slater type atomic orbitals. This was computationally difficult and made no allowance for the effects of electron correlation and so gave unacceptable results. A full NDDO implementation was not possible until 1977 when Dewar and co-workers developed a method for evaluation of these integrals that was both economic and allowed for electron correlation because it was based on an expression which included experimentally derived parameters.

VIII. MODIFIED NEGLECT OF DIATOMIC OVERLAP (MNDO)

In the MNDO approximation the one-center integrals are derived from experimental (atomic) data using the method as in MINDO/3 which continues to allow for the effects of electron correlation. The core–core repulsion, core-electron attraction and resonance integrals are all evaluated by the use of semiempirical expressions:

$$V_{\mu v,B} = -Z_B(\mu^A v^B/S^A S^B) + f_2(R_{AB}) \tag{44}$$

$$\mathbf{CR}_{AB} = Z_A Z_B(S^A S^A/S^B S^B) + f_3(R_{AB}) \tag{45}$$

If $f_2(R_{AB}) = f_3(R_{AB})$ then the core–electron attraction [Eq. (44)] and the core–core repulsion [Eq. (45)] functions nearly cancel. During the parameterization it was assumed that $f_2(R_{AB}) = 0$ and various functions were tried for $f_3(R_{AB})$. The resonance integrals are determined by:

$$\beta_{\mu v} = f_4(R_{AB}S_{\mu v}) \tag{46}$$

The functions f_1–f_4 (f_1 is involved in the expression for the bicentric integrals) were determined by trial and error using a wide variety of functions which satisfied the boundary conditions. After extensive trials,

$$f_3(R_{AB}) = Z_A Z_B(S^A S^A/S^B S^B)[e^{-\alpha_A R_{AB}} + e^{-\alpha_B R_{AB}}] \tag{47}$$

and,

$$f_4(R_{AB}) = \frac{\beta_\mu^A + \beta_\lambda^B}{2} \tag{48}$$

were determined to be the best forms for these functions. The parameters α and β are, in the MNDO approximation, treated as atomic quantities rather than as bond or atom pair quantities, as is the case in MINDO. This causes no loss of accuracy and has the advantage that fewer parameters are required overall. When three or more atoms parameterized, only five additional parameters are required for each additional atom and the atomic parameters are periodic, and cautious extrapolation or interpolation can be applied in the determination of parameters for atoms where experimental data is scarce or unreliable. The parameters α and β were determined by an iterative least squares method which involved full geometry optimization of ten molecules in each of the standard sets for CH, CHN and CHO compounds.

For completeness the expressions for the Fock matrix elements in the MNDO expression are shown below:

$$F_{\mu\mu} = U_{\mu\mu} + \sum_{B} V_{\mu\mu,B} + \sum_{v}^{A} P_{\mu\upsilon}\left[(\mu\mu/vv) - \tfrac{1}{2}(\mu v/\mu v)\right]$$

$$+ \sum_{B}\sum_{\lambda\sigma}^{B} P_{\lambda\sigma}(\mu v/\lambda\sigma) \tag{49}$$

$$F_{\mu v} = \sum_{B} V_{\mu v,B} + \tfrac{1}{2}P_{\mu v}\left[3(\mu v/\mu v) - \tfrac{1}{2}(\mu\mu/vv)\right]$$

$$+ \sum_{B}\sum_{\lambda\sigma}^{B} P_{\lambda\sigma}(\mu v/\lambda\sigma) \tag{50}$$

$$F_{\mu\lambda} = B_{\mu v} - \frac{1}{2}\sum_{v}^{A}\sum_{\sigma}^{B} P_{\mu\sigma}(\mu v/\lambda\sigma) \tag{51}$$

The quantities in the above three expressions can all be evaluated from the semiempirical expressions given previously. This produces a Fock matrix which is a closer approximation to the full Fock matrix than any previous treatment achieved, as it includes all the two-center terms involving monoatomic differential overlap. The result is a method which is superior in its ability to calculate ground state properties (especially those involving angular dependence). Also, the absolute mean error in the calculation of ground state properties with MNDO is only about one half that in the MINDO/3 method, while this is achieved with only a 20% increase in computing time. There are four specific areas of improvement in the MNDO method over earlier schemes. First, the better treatment of aromatic hydrocarbons and triple bonds is more realistic. Because one-center overlap is no longer neglected the treatment for bonds between atoms with unshared lone pairs is improved and, related to this, there are improved bond angles because of the superior treatment of angular dependence in the NDDO approximation. Finally, the energetic ordering of molecular orbitals is much better than in the MINDO/3 method (where spurious high lying σ orbitals were predicted) and the results are in good agreement with photoelectron spectroscopy resulting from a more advanced theoretical description of the difference between π and σ electrons.

MNDO, however, also contained areas of unsatisfactory performance; poor activation energies for many simple reactions, underestimation of the stability for molecules which are "crowded" (e.g., cubane) and an inability to reproduce hydrogen bonds. These problems have been addressed in a "state-of-the-art" NDDO method, that of AM1.

IX. AUSTIN MODEL 1 (AM1)

AM1[21] was introduced by the Dewar group (situated at the University of Texas in Austin) in 1985 after it was recognized that the deficiencies in MNDO resulted from an inappropriate form for the core-repulsion function (Eq. 45). This leads to an overestimation of the repulsion of atoms separated by their Van der Waals distance. In AM1 (so named to avoid any connection with the poorly performing CNDO/INDO based methods) the core-repulsion function has additional Gaussian terms containing adjustable parameters, so,

$$\mathbf{CR}_{AB} = Z_{AB}\gamma_{ss}[1 + F(A) + F(B)] \tag{52}$$

where,

$$F(A) = \exp(-\alpha_a R_{AB}) + \sum_{i}^{4} K_{Ai}\exp[L_{Ai}(R_{AB} - M_{Ai})^2] \tag{53}$$

$$F(B) = \exp(-\alpha_b R_{AB}) + \sum_{j}^{4} K_{Bi}\exp[L_{Bi}(R_{AB} - M_{Bi})^2] \tag{54}$$

The K parameter was found not to be critical and so only the L and M parameters were included for optimization.

Table 1. Comparison of Mean Absolute Errors for AM1, MNDO, and MINDO/3[a]

Average Error In:	*Number of Compounds*	*AM1*	*MNDO*	*MINDO/3*
Heat of formation (kJ mol^{-1})	58	21.21	24.56	40.58
Dipole moments (D)	11	0.17	0.25	0.26
Ionization energies (eV)	22	0.29	0.39	0.31

Note: [a]For hydrocarbons only.[21]

AM1 also represents an improvement over MNDO because of the manner in which the parameters were determined. The optimizable parameters define a "hypersurface" for any given molecular system. The great number of possible values for the parameters make it difficult to determine a set which gives the result which shows the best agreement with observed values. During the development of AM1 a new method allowed a full search of the parameter "hypersurface" to be carried out and so it is believed that the parameters represent a global minimum and possibly the best set for any NDDO based method.

The improvement in performance of AM1 over MNDO and MINDO/3 methods is shown in Table 1.

Other improvements are a correct treatment of hydrogen bonds because of the modified core-repulsion function and better values for activation energies in simple reactions because of the optimal values obtained for the exponents of the Slater type orbitals used in the basis set. These improvements, in particular over MNDO, are achieved without any increase in computation time. This level of accuracy, while retaining economy of calculation means that AM1 is a useful method and closer to a "computational spectrometer" than any previous method.

X. MNDO-PM3 (MNDO-PARAMETRIC METHOD 3)

The most recently developed semiempirical method to date is the PM3 method which was developed by Stewart.[22] The theoretical framework of the PM3 method is identical to that for the AM1 method described earlier, but the development of new methods for optimizing the MNDO parameters combined with the availability of greatly increased computer power together allowed the MNDO parameters to be re-optimized to obtain a set which generally performed better in prediction of all molecular properties than the original set. The PM3 method shows particular improvements in the calculation of the heats of formation of hypervalent compounds: The average error of 106 hypervalent compounds was 56.9 kJ mol^{-1} compared with 317.1 kJ mol^{-1} for MNDO.

The value of the PM3 parameterization has, however, been questioned by Dewar.[23] He has stated his belief that the PM3 method does not provide a sufficient improvement over AM1 to justify its widespread use and distribution and that the comparisons made in relative heats of formation for hypervalent compounds as calculated by each method were erroneous. Stewart[24] has responded to this by stating that the areas in which PM3 performs better than AM1 (namely hypervalent compounds and nitro compounds) justify the use of PM3. He does, however, concede that the method of parameter optimization reveals that the PM3 parameter set, while an improvement on previous sets, is not the global minimum on the parameter hypersurface. Only continued application of both methods to systems outside of those used in the parameterization will reveal the strengths and weaknesses of each.[25]

XI. APPLICATIONS OF SEMIEMPIRICAL METHODS

Each of the modern semiempirical methods (PM3, AM1, MNDO, MINDO/3) have been applied to a wide range of chemical problems, many of which the developers of these methods would not have considered as possible applications. A 1983 review of MINDO/3[17] lists close to 180 references to MINDO/3 calculations of ground state molecular properties as well as physiochemical and spectroscopic parameters for a wide range of chemical and biochemical systems. Furthermore, in the five year period between 1980 and 1985 *Chemical Abstracts* lists 623 papers involving MNDO calculations,[21] and calculations using AM1 have appeared frequently in the literature.

XII. SEMIEMPIRICAL VERSUS AB INITIO: TO WHICH BELONGS THE FUTURE?

In any discussion of semiempirical methods it is inevitable that comparisons to ab initio techniques will be drawn. In fact an objective compari-

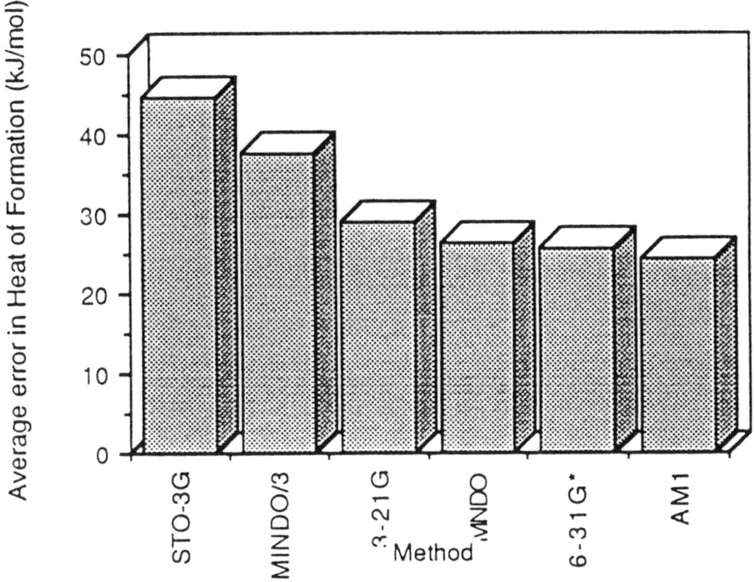

Figure 1. Average errors in heats of formation of 50 molecules calculated by ab initio and semiempirical methods.[26]

son is probably impossible because the two methods judge their success by different criteria. There is no dispute that a semiempirical calculation can be performed much more quickly than a corresponding ab initio one. As computer time attracts substantial charges per hour a semiempirical calculation can represent cost savings, especially on large projects such as the study of potential surfaces. An inaccurate calculation giving misleading results is, however, no substitute for an accurate one regardless of the economic benefits. Dewar has attempted to compare the accuracy of the major modern semiempirical methods with ab initio methods at three levels of theory.[26] A typical result from that investigation—the average error in calculated heats of formation of around 50 small and medium sized molecules—is shown graphically in Figure 1. This indicates that the most recent semiempirical method performs slightly better than the ab initio treatment with a 6-31G* basis set. What is not shown is that it will do the calculations at considerably less cost. An analysis such as this can, however, be misleading. The semiempirical method is parameterized to reproduce heats of formation; it may not perform as well in the calculation of other properties. Lipscomb[27] has pointed out that everyone desires the achievement of accurate and economical calculations, but it must be considered that semiempirical methods, with a heavy reliance on parameters "obscure the physical basis of their success." There are two consequences of this. First, without a rigorously understood theoretical basis the reliability of semiempirical calculations in areas where no comparison with experiment is possible cannot be assessed. A good performance in reproducing ground state properties, for example, does not mean that calculations on the transient species formed during chemical reactions will necessarily be as successful. Second, if the basis for their success lies heavily on parameterization then the development of semiempirical methods does not contribute to any effort to produce more rigorous models that aid in the interpretation of molecular structure. An improvement in the performance generally means an improvement in a set of arbitrary parameters which have limited physical interpretation.

Alternatively, there is a view that is summed up well by a statement of Dewar: "So far as chemistry is concerned the only criterion is practical success."[15] Semiempirical methods have certainly been able to achieve a practical success that makes the arguments against their use, which were discussed above, seem to be overstressing the need for rigorous treatments.

The best approach must be that which views the two methods as complementary. Ab initio methods cannot be satisfactorily applied to all systems of chemical interest, and therefore necessarily some semiempirical treatment is required and in fact is the only alternative. It may be that in the future there will not be a need for semiempirical methods (at least as we know them today), as ab initio methods are being applied with the use of ever faster computers to more and more systems with increasing accuracy and economy. It is also possible that the two methods may begin to converge as new semiempirical methods with fewer and more meaningful parameters are developed.[28] Lindholm[29] has proposed a radical change, moving away from Hartree–Fock theory with its inability to deal with electron correlation, towards a theory based on density functional theory. Although early attempts at implementing these ideas have not been able to achieve a successful parameterization, these ideas have a solid basis in theory.

It is likely that even changes within the the existing MNDO (NDDO) framework may also result in useful improvements in the accuracy of semiempirical calculations. Theil[30] has proposed several refinements to the MNDO type methods including larger basis sets which include d-orbitals, the use of "effective core potentials" and a change in the reference data for energies from heats of formation at 298 K to binding energies. The existing methods will, however, continue to provide information of use to chemists on a wide range of chemical and biochemical problems. If these calculations are carried out with a full awareness of their limitations there seems to be little basis for objection.

REFERENCES AND NOTES

1. Roothaan, C.C.J. *Rev. Mod. Phys.* **1951**, *69*, 23.
2. Hall, G.G. *Proc. R. Soc. London, Ser A.* **1951**, *541*, 205.
3. Parr R.G. *The Quantum Theory of Molecular Electronic Structure*, W.A. Benjamin, NewYork, 1963.
4. McGlynn, S.P.; Vanquickenborne, L.G.; Kinoshita, M.; Carroll, D.G. *Introduction to Applied Quantum Chemistry*, Holt, Reinhart and Wilson, New York, 1972.
5. Hoffmann, R. *J. Chem. Phys.* **1963**, *39*, 1397.
6. Allen, L.C.; Russell, J.D. *J. Chem. Phys.* **1967**, *46*, 1029.
7. Pople, J.A.; Santry, D.P.; Segal, G.A. *J. Chem Phys.* **1965**, *43*, S130.
8. Pople, J.A. *Trans. Faraday Soc.* **1953**, *49*, 1375. Pariser, R.; Parr, R.G. *J. Chem. Phys.* **1953**, *21*, 767.
9. Pariser, R. *J. Chem. Phys.* **1956**, *24*, 250.
10. Pople, J.A.; Segal, G.A. *J. Chem. Phys.* **1965**, *43*, S137.
11. Pople, J.A.; Segal, G.A. *J. Chem. Phys.* **1966**, *44*, 3289.
12. Pople, J.A., Beveridge, D.L.; Dobosh, P.A. *J. Chem. Phys.* **1967**, *47*, 2026.
13. Dixon, R.N. *Mol. Phys.* **1967**, *12*, 83.
14. Baird, N.C.; Dewar, M.J.S. *J. Chem. Phys.* **1969**, *50*, 1262.
15. Dewar, M.J.S., Haselbach, E. *J. Am. Chem. Soc.* **1970**, *92*, 590.
16. Bingham, R.C.; Dewar, M.J.S; Lo, D.H. *J. Am. Chem. Soc.* **1975**, *97*, 1285.
17. Lewis, D.F.V. *Chem. Rev.* **1986**, *86*, 1111.
18. Arenas, J.F.; Qurants, J.J.; Ramirez, F.*J. Theochem.* **1988**, *52*, 143.
19. Dewar, M.J.S. *Science* **1975**, *187*, 1037.
20. Davidson, R.B.; Jorgensen, W.L.; Allen, L.C. *J. Am. Chem. Soc.* **1970**, *92*, 749.
21. Dewar, M.J.S.; Zoebisch, E.G.; Healy, E.F.; Stewart, J.J.P. *J. Am. Chem. Soc.* **1985**, *107*, 3092.
22. Stewart, J.P. *J. Comput. Chem.* **1989**, *10*, 209.
23. Dewar, M.J.S.; Healy, E.F.; Holder, A.J.; Yuan, Y-C. *J. Comput. Chem.* **1990**, *11*, 541.
24. Stewart, J.J.P. *J. Comput. Chem.* **1990**, *11*, 543.

25. A probelm with the core–core repulsion function for PM3 has recently been identified resulting in unrealistic nonbonding potential for hydrogen atoms: Buß, V.; Messinger, J.; Heuser, N. *QCPE Bulletin* **1991**, *5*, 11.
26. Dewar, M.J.S.; Storch, D.M. *J. Am. Chem. Soc.* **1985**, *107*, 3898.
27. Halgren, T.A.; Kleifer, D.A.; Lipscomb, W.N. *Science* **1976**, *190*, 591.
28. Kutzelnigg, W., *Theochem.* **1988**, *181*, 33.
29. Lindholm, E. *Tetrahedron* **1988**, *44*, 7464.
30. Theil, W. *Tetrahedron* **1988**, *44*, 7393.

CONSTRUCTION OF A THREE-DIMENSIONAL MODEL OF THE POLYMERASE DOMAIN OF HIV TYPE 1 REVERSE TRANSCRIPTASE

George R. Painter, C. Webster Andrews,

David W. Barry, and Phillip A. Furman

Advances in Molecular Modeling
Volume 3, pages 47–65.
Copyright © 1995 by JAI Press Inc.
All rights of reproduction in any form reserved.
ISBN: 1-55938-326-7

I. INTRODUCTION

Early after infection of a cell with human immunodeficiency virus (HIV), the two copies of single-stranded genomic RNA contained in the core of the virus are utilized as templates to direct the synthesis of a single copy of double-stranded proviral DNA. The enzymatic activity required to carry out this synthesis resides on a virally encoded Mg^{+2}-dependent reverse transcriptase (RNA-dependent DNA polymerase; RT) also localized in the viral core. The functional RT enzyme purified from virions is heterodimeric, consisting of two subunits of molecular weights 66,000 and 51,000.[1,2] The 66-KDa subunit has both polymerase and RNase H activities. Like other retroviral RTs, the associated RNase H activity is located on the carboxy terminal portion of the polypeptide. Cleavage at a protease-sensitive site on the link between the polymerase and RNase H domains of the 66-KDa polypeptide yields the 51-kDa subunit. The exact function of the 51-kDa subunit is unknown, but it may be involved in allosteric regulation of the enzymatic activity associated with p66 and/or it may be involved in binding the natural $tRNA_3Lys$ primer required to initiate DNA synthesis in the virus.

A bireactant–biproduct mechanism similar to the mechanism described for several other DNA polymerases has been proposed for DNA synthesis catalyzed by RT.[3] A simplified reaction scheme is shown in Figure 1. Binding of reactants is ordered in this mechanism with the template-primer (TP^N) binding first followed by a 2′-deoxynucleoside 5′-triphosphate (dNTP). If the dNTP has Watson–Crick complementarity to the next appropriate template residue, the ternary ground state complex is proposed to undergo a conformational change to form the catalytically competent complex. Incorporation of the bound dNTP into the nascent chain then occurs via an S_N2 displacement of pyrophosphate (PP_i) by the 3′-hydroxyl at the primer terminus.[4]

Catalytically active HIV-1 RT has been cloned and expressed in *Escherichia coli*.[5] The recombinant heterodimeric enzyme is physically

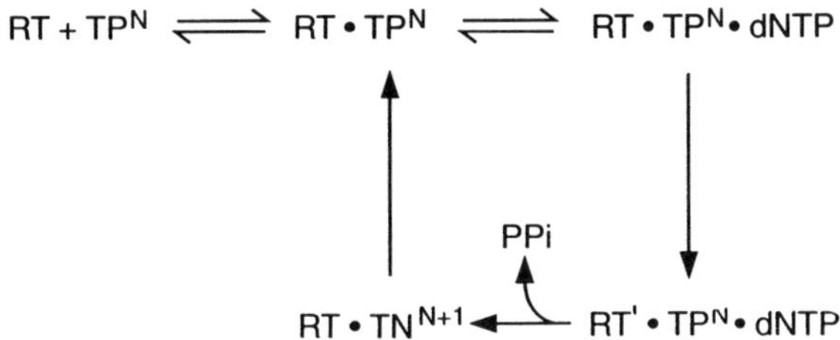

Figure 1. The catalytic cycle of HIV-1 reverse transcriptase (RT). Binding of the reactants is ordered with template-primer (TP^N, where N denotes the number of nucleotides in the primer strand) binding first followed by a 2′-deoxynucleoside 5′-triphosphate (dNTP). If the dNTP has Watson–Crick complementarity to the next appropriate template residue, the ternary ground state complex (RT·TP^N·dNTP) undergoes a conformational change to form the catalytically competent complex (RT′·TP^N·dNTP). Incorporation of the 5′-monophosphate of the bound dNTP into the nascent chain then occurs by an S_N2 displacement of pyrophosphate (PP$_i$) by the 3′-hydroxyl at the primer terminus.

and kinetically similar to the native enzyme purified from virus. It should be noted that although the RT is the only enzyme required, in principle, to generate double stranded DNA, synthesis of full-length viral DNA can only be carried out when the RT is contained in or associated with an intact viral core. Consequently, a true understanding of the factors affecting the catalytic efficiency, fidelity, and substrate specificity of this enzyme may ultimately require studying the RT in the context of the viral core.

The availability of recombinant HIV-1 RT in 1987 made it possible to begin to probe for sites important in reactant recognition and binding, catalytic function and the development of resistance to antiviral drugs. These functional studies would all have been facilitated by some knowledge of the three-dimensional structure of the RT, particularly of the dNTP binding site. However, very little information was available on the topology of the dNTP or TP binding sites at the time these studies were initiated. Until very recently, attempts to obtain X-ray diffraction data at high resolution were unsuccessful. Because of the pressing need to establish structure–function relationships to aid in the design of new anti-HIV drugs and to help in developing an understanding of the biochemical basis of drug resistance, we decided to construct a three-dimensional model of the active polymerase domain of HIV-1 RT using comparative protein model building techniques. The primary purpose of the model was to help in interpreting the results of biochemical and biophysical experiments until such time as a high resolution X-ray structure became available. It had been demonstrated that, in certain case, three-dimensional models of proteins of unknown structure could be built based on functional homology and sequence similarity to proteins of known three-dimensional structure.[6–10] This type of model had proven to be very useful in the planning and interpretation of biochemical and mutagenesis experiments.[11–13]

Here, we describe the construction of this model using the Klenow fragment of *E. coli* DNA polymerase I as a starting point, and compare the model to the recently published structure of the polymerase domain of HIV-1 RT determined using X-ray crystallography.[14] Of particular interest in terms of the function of the enzyme was the shape and charge profile of the nucleic acid binding cleft and the predicted location of the dNTP binding site derived from mapping site mutations associated with drug resistance (AZT, ddI and ddC) onto the model.

II. COMPARATIVE PROTEIN MODEL BUILDING

A. Location of Structurally Conserved Regions (SCRs)

Comparative protein model building utilizes spatial information from one or more known structures within a family of proteins to construct a model of a member of the family with an unknown structure.[15,16] At the time we started building the model, the Klenow fragment of *E. coli* DNA polymerase I was the only polymerase-containing structure known at atomic resolution (α carbons only),[17] therefore the model building was necessarily based on a stringent comparison of the primary and secondary structures of the polymerase domains of the Klenow fragment and HIV-1 RT.

The first step in the construction of such a model is to compare the amino acid sequences of the two proteins and to locate areas of similar amino acid composition. Sequence alignments of the polymerase domains of HIV-1 RT and the Klenow fragment were carried out using both the ALIGN_SEQUENCE command in SYBYL Version 5.4 (Tripos

Associates, St. Louis, MO) and the GAP procedure in the Protein Comparison Module of the sequence analysis software package of the University of Wisconsin Genetics Computer Group (UWGCG) Version 6.0. SYBYL was run on a Silicon Graphics Personal Iris and the UWGCG program on a VAX 11/780. Both programs use the Needleman–Wunsch algorithm and a Dayhoff conservative mutation matrix.[18,19] The portion of the primary sequence of HIV-1 RT (HXB2 strain) used for the alignment consists of 424 residues beginning at P156, the amino terminus of the polymerase domain,[20] and extending to K579, which is adjacent to a protease sensitive site on the linker between the polymerase and RNase H domains.[21] For purposes of constructing the model, the HIV-1 RT sequence was renumbered from 1 (P156) to 424 (see Figure 2).

The portion of the primary sequence of Klenow used for the alignment extends from K518 to H928 (the 411 residue fragment that contains the polymerase activity of the enzyme).[17,22] The two sequences were aligned with as few gaps as possible consistent with obtaining a good alignment. This approach to primary structure comparison is based on the assumption that the proteins are related principally by missense substitution of amino acids and emphasizes their structural and evolutionary relatedness.[23,24] Some manual adjustments were made to the alignment based on secondary structure considerations.

When aligning the primary sequences of the polymerase domains of HIV-1 RT and the Klenow fragment, it was initially assumed that those portions of the polypeptide chain essential to catalysis were not transposed. Given the functional similarity of the two enzymes and the fact that the two polymerase domains are of almost identical size, the probability of such a rearrangement was thought to be low. The alignment of the two sequences made with this assumption gives an identity ranging from 22 to 34% and a similarity ranging from 50 to 61% depending on the gap parameters used in the sequence alignment algorithm. It was subsequently suggested that a transposition of motifs had occurred, and that incorporation of these transpositions when aligning the two sequences would result in a more significant apparent homology.[25] Selected portions of the alignment in which the transpositions were included are shown in Figure 2. Models were built using both straight and transposed alignments.

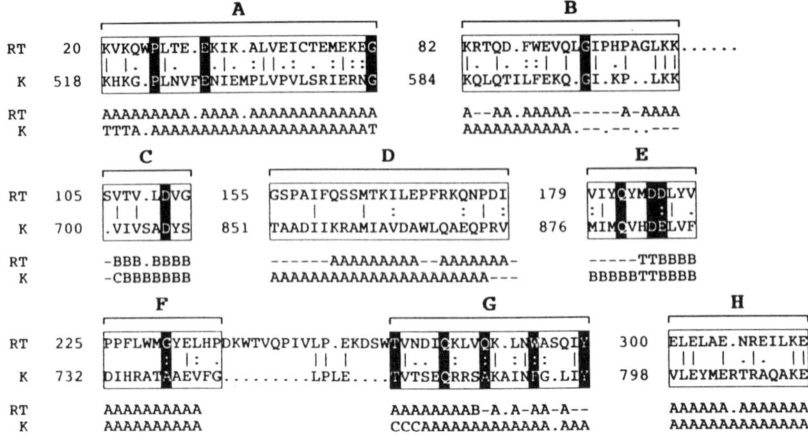

Figure 2. Amino acid sequence alignment of portions of the polymerase domains of HIV-1 RT (RT) and Klenow (K). The sequences are numbered consecutively from the amino to the carboxy terminus. The SCRs are enclosed in boxes and are lettered A through H. HIV-1 RT residues that are highly conserved in the polymerase family,[31] and are aligned with identical or conservatively mutated residues in the Klenow sequence are enclosed in black boxes. The positions of secondary structural elements identified in the X-ray structure of the Klenow fragment and predicted for HIV-1 RT are shown on the two lines below the aligned sequences. α-Helical residues are denoted by A, residues in a β-sheet are indicated by B, residues in a random coil are denoted by C, and residues in a turn are denoted by T.

The location of conserved secondary structural elements (α-helices, β-sheets, etc.) can also be used when comparing proteins in the same family. Secondary structure analysis was carried out using the UWGCG sequence analysis program Peptide Structure and the PREDICT_SEC-ONDARY_STRUCTURE command in SYBYL. Predictions made with the UWGCG program use the Chou–Fasman and Garnier–Osguthorpe–Robson methods,[26,27] and the predictions made in the SYBYL program (BAYES STATISTICS) use the method of Maxfield and Scheraga.[28] The consensus secondary structure for selected portions of the HIV-1 RT polymerase produced by addition of the individual secondary structure probabilities at each position is shown (Figure 2) along with the secondary structure of selected portions of the polymerase domain of the

Klenow fragment. The predicted secondary structure of HIV-1 RT is in good agreement with that previously published for the first 328 residues of the sequence.[24] The secondary structure shown for the Klenow sequence was taken from the X-ray crystal structure with the exception of the segment between K584 and K601, which was predicted. Due to local crystal disorder in the polymerase domain of the Klenow fragment, coordinates for two segments of the protein backbone, A574 to P622 and Q780 to Y787, could not be obtained.[17]

Areas having high local sequence homology and/or comparable secondary structure were taken to represent functionally significant portions of the framework, and were used as SCRs for model building. A similar approach was utilized by Pearl and Taylor to construct a model of the retroviral proteases from structural data on aspartic proteases.[8] The SCRs, shown enclosed in boxes in Figure 2, represent those portions of the HIV-1 RT sequence with the highest primary and secondary structural similarity to the Klenow sequence. Included in the SCRs are 125 of the 424 residues in the sequence of the polymerase domain of HIV-1 RT. Insertions and deletions were kept to a minimum in positioning the SCRs. Those that remained were accommodated as previously described by Greer.[29,30]

Primary and secondary structure comparisons reported on RNA- and DNA-dependent RNA and DNA polymerases support the selection of SCRs shown in Figure 2. Alignments of the amino terminal segments of numerous retroviral RTs, including sequences from the lentivirus sub-family, have shown significant sequence conservation beginning at amino acid 20 and ending around amino acid 190.[24,31] Barber et al. have proposed that this represents evolutionary conservation and is the result of the vital role the first 190 amino acids play in RT function, possibly forming the inner surface of a cleft within which nucleic acid TPs bind.[24] According to the modeling studies of Ollis et al., the aligned portion of the Klenow fragment extending from lysine 518 to valine 701 (Figure 2) forms a surface that contacts bound duplex B-DNA.[17] Thus, the identification of five SCRs (A, B, C, D, and E) in the first 190 residues of the HIV-1 RT sequence is not surprising. Furthermore, the α-helices proposed for SCRs A, F, and G are consistent with the consensus secondary structural predictions previously made for ten aligned retroviral RT sequences.[24] The homologous sequences aligned to form SCR G have been proposed to be a common structural element within the polymerase family.[32] The equivalence of the HIV-1 RT and Klenow sequences aligned in SCR G has previously been suggested based on the results of chemical modification studies.[33]

B. Construction of the Model of the Polymerase Domain

Model building was performed on a Silicon Graphics Personal Iris using the COMPOSER[16] module of SYBYL. The three-dimensional template used for the placement of SCRs identified in the polymerase domain of HIV-1 RT was the 2.8 Å α-carbon map of the Klenow fragment of *E. coli* DNA polymerase I.[17] The atomic coordinates of the backbone α-carbons of the Klenow fragment were obtained from the Brookhaven Protein Data Bank, Brookhaven, NY (1DPI).[34] An all-atom model of Klenow was created using the CONSTRUCT-BACKBONE/ADD-SIDECHAIN commands in SYBYL. Since side chain positions in the Klenow fragment were not experimentally determined,[17] they were positioned using a rule-based approach in which the most commonly observed rotamer for each residue type is used.[35]

The following procedure was used to build the model of the polymerase domain of RT: (1) Having established the positions of the RT SCRs on the Klenow structure via primary sequence alignment and secondary structure analysis, the HIV-1 RT residues were assigned the coordinates of the corresponding Klenow residues using the MODEL_CONSERVED_REGION command of COMPOSER. (2) Regions not defined in step 1 were built-in using LOOP SEARCH. This procedure searches a protein database containing high resolution structures from the Protein Data Bank to find fragments of the appropriate length and geometry. Appropriate loops were chosen based on structural and sequence criteria as previously described by Giranda et al.[9] Loop structures that interfered sterically with the remainder of the structure were eliminated first. From the remaining choices, the loop with primary and secondary structural similarity to the HIV-1 sequence under consideration was chosen. Once again, side chain positions were added using a rule-based approach. (3) An all-atom model was generated, standard AMBER charges assigned and the system minimized in two phases using the AMBER force field,[36] a 7 Å nonbonded cutoff, and a distance dependent dielectric.

In phase one of the minimization procedure, the backbone atoms were constrained with a 1000 kcal/mole harmonic constraint and the sidechains minimized until the rms gradient fell below 1.0 kcal/Å. In

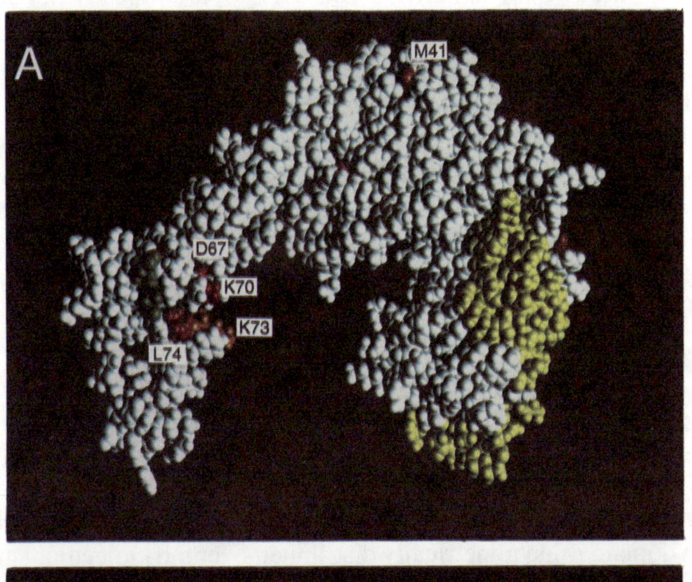

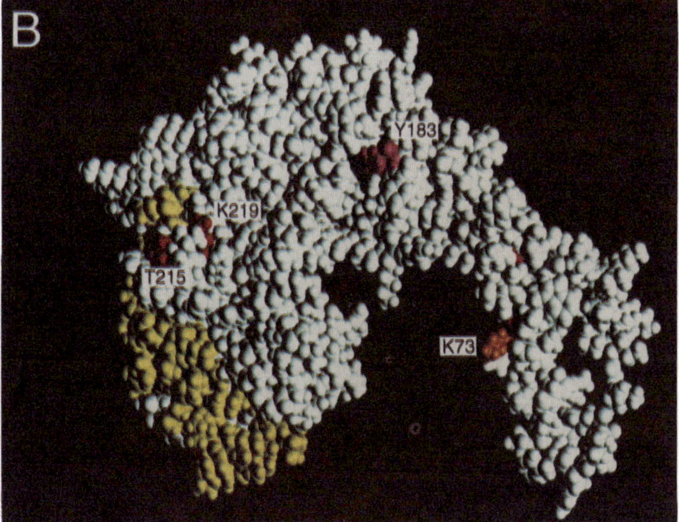

Figure 3. Space filling representations of the polymerase domain viewed from (A) the amino terminus and (B) the carboxy terminus. The positions of amino acids and sequences of amino acids that have been associated with specific functions are shown mapped onto the polymerase model. Residues implicated in the development of resistance to antiviral dideoxynucleosides are labeled and shown in red. The putative dimerization site, L283 to G335, is shown in yellow. Y183, a residue that is highly conserved in most RNA and DNA polymerases is labeled and shown in purple. The site of dTTP crosslinking, K73, is shown in orange.

phase two, backbone constraints were removed and a full minimization was done until the rms gradient fell below 0.1 kcal/Å. The minimized model is shown in Figures 3A and 3B. Each of the first five SCRs, which are in the segment of the primary sequence proposed to be involved in nucleic acid binding,[24] lie on the endoface of the proposed cleft. The dimensions of the cleft taken from the model are approximately 23 Å wide, 20 Å deep, and 57 Å long on the longest side.

C. Calculation of the Electrostatic Charge Potential

An electrostatic potential grid calculation was carried out in SYBYL using AMBER charges, a distance dependent dielectric, a 100 Å cutoff, and a grid-spacing of 1.0 Å. There are 57 acidic and 62 basic residues within the polymerase domain of HIV-1 RT. The excess positive charge on the domain is asymmetrically distributed over the molecular surface. The positive potential is located predominantly on one side of the molecule, the side that forms the endoface of the cleft, with the highest positive electrostatic charge density concentrated along the sides of the cleft adjacent to the α-helical segment extending from K64 to K73, and on the loop between K276 and K282. The areas of negative electrostatic potential generally lie outside of the cleft with two critical exceptions: (1) the antiparallel β-hairpin structure containing the highly conserved Y183–M184–D185–D186 sequence,[37,38] and (2) the adjacent β-strand containing D110 and D113. The negative electrostatic potential on these two adjacent segments of the protein is shown in Figure 4. The overall distribution of charge on the entire polymerase domain and the shape of the calculated potential grid are comparable to those of the Klenow fragment.[17]

The Y183–M184–D185–D186 region (see Figure 4) has been proposed to play an essential role in the catalysis of the polymerase reaction. D185, D186, and D110 are positioned properly for binding divalent metal ions. Divalent cations such as Mg^{+2} or Mn^{+2} are required cofactors for RNA and DNA polymerases.[39] Binding of dNTP substrates to polymerases has been proposed to involve formation of an enzyme–Mg^{+2} substrate bridge.[40] The Mg^{+2} is believed to bind to the substrate by inner sphere coordination to the γ-phosphoryl group and second sphere coordination to the α- and β-phosphoryl groups. Binding of Mg^{+2} to the 5′-triphosphate moiety of the dNTP may also activate pyrophosphate (PP_i) as a leaving group.

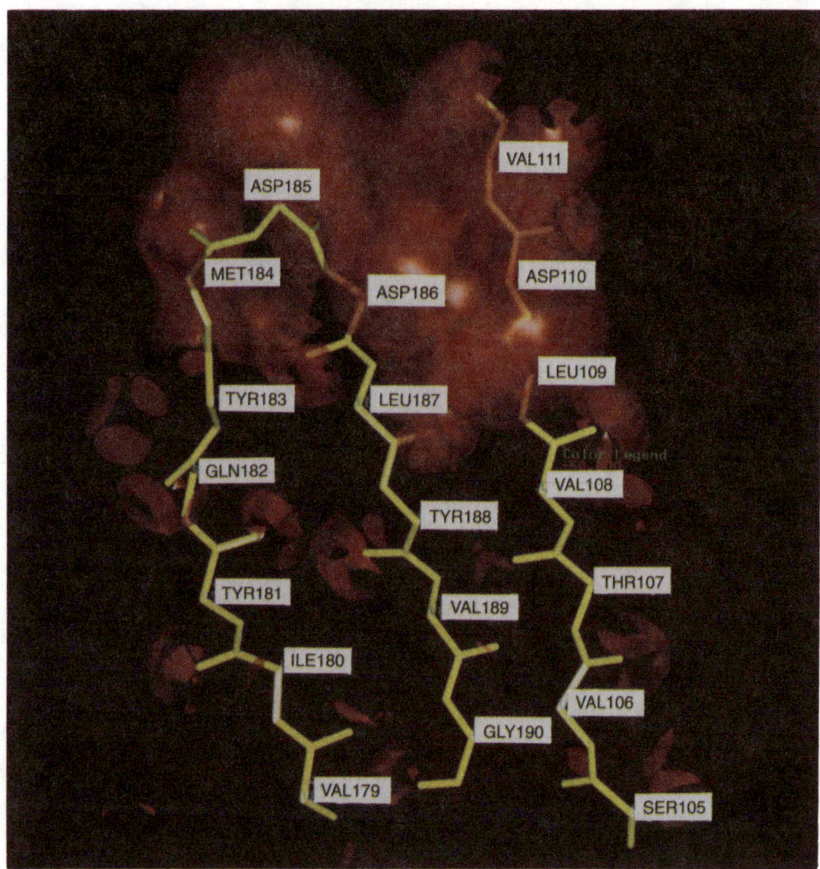

Figure 4. A stick representation of the β-hairpin structure containing the YMDD sequence (SCR E) and an adjacent β-strand containing D110 (SCR C). The negative charge potential on this portion of the protein (shown in red) is contoured at –50 kcal/mol.

III. STRUCTURE–FUNCTION RELATIONSHIPS BASED ON THE MODEL

A. The Nucleic Acid Binding Cleft

A key function of the polymerase domain of HIV-1 RT is to bind the nucleic acid TPs (either RNA or DNA) that select via Watson–Crick complementarity the 2′-deoxynucleoside 5′-monophosphates that are incorporated into the nascent DNA strand. The model presented here has the topological and physicochemical features necessary to carry out this function. The size of the cleft in the minimized model (Figures 3A and 3B) is appropriate for binding the polyanionic template-primers. In addition, the electrostatic potential surface generated from the model is asymmetric with a strong positive potential well existing within the cleft, and regions of strong negative potential on the outer surface (see section IIC). This type of potential surface is observed for other nucleic acid binding proteins. Calculations of the electrostatic charge potentials of several DNA and RNA binding proteins have shown that the electrostatic fields around the proteins are of the appropriate charge and shape for facilitating the orientation and docking of the polyanionic, nucleic acid substrates.[41] In the case of Klenow, positive potential is localized within the DNA binding cleft, while centers of negative potential are positioned outside of the cleft.[17] This distribution could effectively reduce the number of electrostatically nonproductive complexes and assist in the proper association geometry with DNA. In addition, the potential may have, as has been suggested for calmodulin,[42] a conformational restriction role.

Some degree of flexibility in the TP binding region may be necessary to produce shape complementarity within the TP and thereby ensure a relatively high binding affinity (10^{-8} to 10^{-9} M^{-1},[43]) with the wide array of three-dimensional structures encountered when utilizing various species of nucleic acid as templates. The limited amount of molecular dynamics carried out on the model demonstrated that although a majority of the structure stayed intact over the time trajectory computed (50 ps), some movement was observed in the structure between residues 78 and 103. This segment of the molecule, which constitutes an important part of the nucleic acid binding domain of the polymerase, corresponds to the portion of the Klenow molecule that was not resolved in the X-ray analysis as a result of high thermal variability.[17] Along with the changes in conformation in this flexible region, there are concomitant changes in the electrostatic potential. The changes in the surface potential may also help to modulate the affinity of the enzyme for various TPs.

B. Location of the Substrate Binding Site

dNTP binding experiments using intrinsic protein fluorescence, and photoaffinity labeling using [α-^{32}P]dTTP as a reactant have shown that the dNTPs and AZTTP compete for a single site or set of overlapping sites on the p66 subunit of the heterodimeric enzyme.[43,44] Transferred nOe experiments carried out on dTTP-, AZTTP-, and dATP-RT complexes confirm the existence of a set of overlapping dNTP binding sites that contain common contact points on the protein surface (Dr. Ann Aulabaugh, *Biochem. Biophys. Acta*, in press). In the case of dTTP, the photoaffinity labeling experiments have indicated one contact point to be K73 (Dr. Nancy Cheng, *Biochemistry*, in press). The position of K73 and other amino acid residues or sequences of amino acids implicated in specific functions of the polymerase as revealed by site directed mutagenesis are shown mapped onto the polymerase model in Figures 3A and 3B. Mutation sites associated with the development of resistance to dideoxynucleoside type drugs are shown in red. One set of these mutation sites (D67 and K70[45]; L74[46]) is on an α-helical segment that lies on the endoface of the cleft, and surrounds the dTTP crosslinking site at K73 (shown in orange in Figure 3A). The spatial relationship of these residues strongly suggests this portion of the structure may constitute a part of the dNTP binding site(s).

Although the precise biochemical mechanism for the induction of resistance in HIV-1 to dideoxynucleoside drugs by mutations within the polymerase domain is at present unknown, work done on the origins of the replication fidelity exhibited by polymerases suggests several possibilities.[47,48] Of these, the most straightforward explanation is that a direct interaction occurs between the dNTP and a specific residue at the binding site. Alteration of that residue results in a less energetically favorable interaction with the binding site, an increased K_d for the analogue and a proportional increase in K_i. Alternatively, the site mutation may affect the ability of the α-phosphoryl moiety of the dNTP to assume the appropriate orientation in the trigonal–bipyridal transition state formed in the chain elongation step. These are the two most likely explanations for the effects of the mutations clustered around the putative dNTP binding site.

In the model and the recently published X-ray structure of RT,[14] the mutations clustered around K73 are distal to D185, D186, and D110. A model of the template–primer–RT complex constructed using the X-ray structure has the primer terminus adjacent to these three residues, suggesting a critical role for them analogous to the role of the corresponding residues in Klenow, in catalysis.[14] As a consequence, it was considered unlikely that the mutations around K73 constitute part of the dNTP binding site and therefore do not affect direct interactions between the protein and the dideoxynucleotide inhibitors. However, this conclusion is hard to rationalize in light of the fact that dTTP crosslinks exclusively to K73.

There are additional mutations associated with the development of drug resistance that are distal to the proposed dNTP binding site.[45,46] M41, mutation of which results in resistance to AZT, is located on the upper edge of the amino terminal face of the cleft (Figure 3A), while T215 and K219 are located on the back edge of the cleft (Figure 3B). Mutation of T215 to Y and K219 to Q also result in the development of resistance to AZT. Changes at these sites may either induce changes in the conformation of the enzyme that in turn alter the configuration of the substrate recognition site, or these residues may be involved in overall base pair recognition during the final states of dNTP incorporation. It has been proposed that the chain elongation reaction is driven forward, and appropriate dNTP selection enhanced, by exclusion of water around base pairs in the active site cleft of the enzyme.[49] An amino acid mutation could alter the architecture of the cleft such that the dehydration that accompanies base stacking and Watson–Crick hydrogen bonding cannot proceed effectively when the incoming dNTP has been modified.

C. p66/p51 Dimer Interface

The HIV-1 RT isolated from virions is a p66/p51 heterodimer.[1] Photoaffinity labeling of the heterodimer using $[^{32}P]rA_{12-18}{\cdot}dT_{10}$ as a representative TP and $[\alpha\text{-}^{32}P]dTTP$ as a representative dNTP results in crosslinking exclusively to the p66 subunit.[44] These data suggested that the active polymerase domain resides on the p66 subunit, and the polymerase domain of the p51 subunit must be in a significantly different conformation since the crosslinking sites, which the model indicated to be within the cleft, are no longer accessible to the reagents. This conclusion was ultimately borne out by the X-ray crystal structure. Available data also suggested that the dimer interface was hydrophobic and was located toward the carboxy terminus of the polymerase domain, possibly between L283 and G335.[31] This portion of the model, in accord with this function, is located on the outside surface of the protein and has a high solvent accessibility (see Figures 3A and 3B, yellow residues). The actual interaction between the subunits of the heterodimer, as revealed by X-ray analysis, was much more complex and more extensive than anticipated from site mutation data. Given the biochemical information available at the time, there was no way modeling could have been used to predict the actual mode of monomer interaction.

IV. EVALUATION OF THE MODEL AND DERIVED STRUCTURE–FUNCTION RELATIONSHIPS

A. Use of Klenow for Comparative Model Building

Why did we believe that a usable model of the active polymerase domain of HIV-1 RT could be built using only Klenow? Even though building a model based on a single example is tenuous, comparative modeling has been demonstrated to work in cases where only one reference structure has been used,[50,51] and in cases where sequence homology to the single reference protein is low.[52] In the case of HIV-1 RT and the Klenow fragment of *E. coli* DNA polymerase I, there was compelling structural and biochemical data that led us to believe such a comparison could provide a usable structure. Structural comparisons of proteins that perform similar functions had demonstrated that three-dimensional structure can be conserved even in instances when sequence homology was somewhat low, such as with hen egg white lysozyme and T4 bacteriophage lysozyme.[51] The suggestion of a common fold for the polymerase family had already been made based on comparisons of the primary structures of numerous RNA- and DNA-directed RNA and DNA polymerases.[38,53] In a global sense, this prediction was demonstrated to be true in the case of HIV-1 RT and Klenow. X-ray crystallographic analysis had shown the Klenow fragment to have a deep crevice of the appropriate size and shape for binding nucleic acid template-primers.[17,54] A crevice of similar size and shape had been seen in electron micrographs of HIV-1 RT.[55] Single- and double-stranded DNA appeared to bind in this crevice. The fact that the fold used to produce such a crevice may be general to the polymerase family was further demonstrated by the report of the shape of the *E. coli* RNA polymerase holoenzyme determined by electron crystallography.[56]

Klenow and HIV-RT are also functionally similar. Both carry out DNA directed DNA synthesis via an ordered reaction mechanism.[3] Furthermore, the molecular forces that dictate the affinity and specificity shown by both enzymes for nucleic acid and nucleotide reactants are similar.[43,57] Thus, structurally and functionally, the polymerase domain of the Klenow fragment appeared to be an extremely good starting point for modeling the polymerase domain of HIV-1 RT, and the comparative modeling process was carried forward despite the fact there was only one known structure available.

B. Comparison of the Model with X-ray Crystal Structure

How accurate was the polymerase model produced using this process? Comparison is best made on the subdomain level. Due to the anatomical resemblance of the active polymerase domain to a right hand, Steitz and co-workers have named the subdomains of this segment of the enzyme as "finger," "palm," and "thumb."[14] The subdomain in which the correspondence between the model and the X-ray structure is greatest is the "palm," which contains the putative catalytic site. A portion of the "palm" is shown in Figure 4. This region of the model consists of five β-strands and two α-helices that are superimposable on the corresponding secondary structural elements in the "palm" subdomain of the X-ray structure. This result is consistent with the idea that all the polymerases are related evolutionarily, and that the region most likely to be highly conserved is that containing residues involved in catalysis. The position of the "palm" subdomain in the model (indicated by Y183 in Figure 3B) on the roof of the nucleic acid binding cleft is comparable to the position of the "palm" in the X-ray structure.

The correspondence between the remainder of the model and the X-ray structure is generally poor. The "thumb" domain of HIV-1 RT consists of a bundle of four α-helices, while the "thumb" of the model consists of two extended helices. Furthermore, the alignment (Figure 2) in this portion of the model is inaccurate. The "finger" subdomains of HIV-1 RT and the Klenow fragment are considerably different. Consequently, Klenow was not a good template for construction of this subdomain. The "fingers" region of HIV-1 RT is predominantly β-sheet, while that of Klenow is predominantly α-helix.

Because the polymerase model was built using a single example, we anticipated that the overall accuracy of the structure would probably be low. Therefore, we felt that the most appropriate use of the model would be to establish gross spatial relationships between groups of amino acid residues involved in catalysis, substrate binding, and drug resistance. These low resolution structure–activity relationships would then be used in interpreting data derived from mechanistic and site mutation studies, and to design additional site mutation studies. A specific case in which the model was used in this fashion involves the mutations of T215, K219, D67, K70, and L74 that are involved in the development of resistance to dideoxynucleoside drugs. It had been suggested that the fold of the protein may be such that all of these mutations are proximal and that they may constitute part of the dNTP binding site. However, the model showed the mutations at 215 and 219 to be on the opposite side of the template-primer binding cleft from the mutations clustered around K70 (see Figures 3A and 3B). This result was verified in the X-ray crystal study. The significance of this prediction was that the mechanism by which these two groups of mutations induce drug resistance must be different (Section IIIB). This is the level at which the model proved useful, and provided accurate insight into general spatial relationships within the polymerase domain. Any higher resolution use of the model, such as for docking substrates, would have been inappropriate.

In conclusion, the comparative model building approach provided a low resolution model of the active polymerase domain of HIV-1 RT. This model was used to interpret the results of mechanistic studies of the enzyme to establish spatial relationships within the polymerase domain, and to study the binding site of the nonnucleoside inhibitor nevirapine.[58]

REFERENCES

1. Hoffman, A.D.; Banapour, B.; Levy, J.A. *Virology* **1985**, *147*, 326–335.
2. DiMarzo, V.T.; Copeland, T.D.; DeVico, A.L.; Rahman, R.; Orozlan, S.; Gallo, R.C.; Sarngadharan, M.G. *Science* **1986**, *231*, 1289–1291.
3. Majumdar, C.; Abbotts, J.; Broder, S.; Wilson, S.H. *J. Biol. Chem.* **1988**, *263*, 15657–15665.
4. Hopkins, S.; Furman, P.A.; Painter, G.R. *Biochem. Biophys. Res. Commun.* **1989**, *163*, 106–110.
5. Larder, B.; Purifoy, D.; Powell, K.; Darby, G. *EMBO J.* **1987a**, *6*, 3133–3137.
6. Blundell, T.; Sibanda, B.L.; Pearl, L. *Nature* **1983**, *304*, 273–275.
7. Ripka, W.C. *Nature* **1986**, *321*, 93–94.
8. Pearl, L.H.; Taylor, W.R. *Nature* **1987**, *329*, 351–354.
9. Giranda, V.L.; Chapman, M.S.; Rossmann, M.G. *Proteins* **1990**, *7*, 227–233.

10. Weber, P.C.; Lukas, T.J.; Craig, T.A.; Wilson, E.; King, M.M.; Kwiatkowski, A.P.; Watterson, D.M. *Proteins* **1989**, *6*, 70–85.

11. Burnbaum, J.J.; Starzyk, R.M.; Schimmel, P. *Proteins* **1990**, *7*, 99–111.

12. Struthers, R.S.; Kitson, D.H.; Hagler, A.T. *Proteins* **1991**, *9*, 1–11.

13. Haffey, M.L.; Novotny, J.; Bruccoleri, R.E.; Carroll, R.D.; Stevens, J.T.; Matthews, J.T. *J. Virology* **1990**, *64*, 5008–5018.

14. Kohlstaedt, L.A.; Wang, J.; Friedman, J.M.; Rice, P.A.; Steitz, T.A. *Science* **1992**, *256*, 1783–1790.

15. Browne, W.J.; North, A.C.T.; Phillips, D.C.; Brew, K.; Vanaman, T.C.; Hill, R.L. *J. Mol. Biol.* **1969**, *42*, 65–86.

16. Blundell, T.; Carney, D.; Gardner, S.; Hayes, F.; Howlin, B.; Hubbard, T.; Overington, D.; Singh, D.A.; Sibanda, B.L.; Sutcliffe, M. *Eur. J. Biochem.* **1988**, *172*, 513–520.

17. Ollis, D.L.; Brick, P.; Hamlen, R.; Xuong, N.G.; Steitz, T.A. *Nature* **1985**, *313*, 762–766.

18. Needleman, S.B.; Wunsch, C.D. *J. Mol. Biol.* **1970**, *48*, 443–453.

19. Dayhoff, M.O.; Schwartz, R.M.; Orcutt, B.C. In *Atlas of Protein Sequence and Structure*, vol. 5, Dayhoff, M.O.; Schwartz, R.M.; Orcutt, B.C. (eds.), National Biomedical Research Foundation, Washington, DC, 1978, 345–352.

20. Ratner, L.; Haseltine, W.; Patarca, R.; Livak, K.J.; Starcich, B.; Josephs, S.F.; Doran, E.R.; Rafalski, J.A.; Whitehorn, E.A.; Baumeister, K.; Ivanoff, L.; Petteway, S.R., Jr.; Pearson, M.L.; Lautenberger, J.A.; Papas, T.S.; Ghrayeb, J.; Chang, N.T.; Gallo, R.C.; Wong-Staal, F. *Nature* **1985**, *313*, 277–284.

21. Lowe, D.M.; Aitken, A.; Bradley, C.; Darby, G.K.; Larder, B.A.; Powell, K.L.; Purifoy, D.J.M.; Tisdale, M.; Stammers, D.K. *Biochemistry* **1988**, *27*, 8884–8889.

22. Freemont, P.S.; Ollis, D.L.; Steitz, T.A.; Joyce, C.M. *Proteins* **1986**, *1*, 66–73.

23. Doolittle, R.F. *Science* **1981**, *214*, 149–159.

24. Barber, A.M.; Hizi, A.; Maizel, J.V., Jr.; Hughes, S.H. *AIDS Res. Human Retroviruses* **1990**, *6*, 1061–1072.

25. Taylor, E.W.; Jaakkala, J. *Genetica* **1991**, *84*, 77–86.

26. Chou, P.Y.; Fasman, G.D. *Adv. Enzymol.* **1978**, *47*, 45–148.

27. Garnier, J.; Osguthorpe, D.J.; Robson, B. *J. Mol. Biol.* **1978**, *120*, 97–120.

28. Maxfield, F.R.; Scheraga, H.A. *Biochemistry* **1976**, *15*, 5138–5153.

29. Greer, J. *J. Mol. Biol.* **1981**, *153*, 1027–1042.

30. Greer, J. *Proteins* **1990**, *7*, 317–334.

31. Jacobo-Molina, A.; Arnold, E. *Biochemistry* **1991**, *30*, 6351–6361.

32. Delarue, M.; Poch, O.; Tordo, N.; Moras, D.; Argos, P. *Protein Eng.* **1990**, *6*, 461–467.

33. Basu, A.; Tirumalai, R.S.; Modak, M.J. *J. Biol. Chem.* **1989**, *264*, 8746–8752.

34. Bernstein, F.C.; Koetzle, T.F.; Williams, G.J.B.; Meyer, E.F., Jr.; Brice, M.D.; Rogers, J.R.; Kennard, O.; Shimanouchi, T.; Tasumi, M. *J. Mol. Biol.* **1977**, *112*, 535–542.

35. Suttcliffe, M.J.; Hayes, F.R.F.; Blundell, T.L. *Protein Eng.* **1987**, *1*, 385–392.

36. Weiner, S.; Kollman, P.; Nguyen, D.; Case, D. *J. Comp. Chem.* **1986**, *7*, 230.

37. Argos, P. *Nucleic Acids Res.* **1988**, *16*, 9909–9916.

38. Kamer, G.; Argos, P. *Nucleic Acids Res.* **1984**, *12*, 7269–7281.

39. Mildvan, A.S.; Loeb, L.A. *CRC Critical Reviews Biochem.* **1979**, *6*(2), 219–244.

40. Sloan, D.L.; Loeb, L.A.; Mildvan, A.S.; Feldman, R.J. *J. Biol. Chem.* **1975**, *250*, 8913–8919.

41. Matthew, J.B. *Ann. Rev. Biophys. Chem.* **1985**, *14*, 387–417.

42. Weber, I.T.; Miller M.; Jaskolski, M.; Leis, J.; Skalka, A.M.; Wlodawer, A. *Science* **1989**, *243*, 928–931.

43. Painter, G.R.; Wright, L.L.; Hopkins, S.; Furman, P.A. *J. Biol. Chem.* **1991**, *266*, 19362–19368.

44. Cheng, N.; Painter, R.R.; Furman, P.A. *Biochem. Biophys. Res. Commun.* **1991**, *174*, 785–789.

45. Larder, B.A.; Kemp, S.D. *Science* **1989**, *246*, 1155.

46. St. Clair, M.H.; Martin, J.L.; Tudor-Williams, G.; Bach, M.C.; Vavro, C.L.; King, D.M.; Kellam, P.; Kemp, S.D.; Larder, B.A. *Science* **1991**, *253*, 1557–1559.

47. Carroll, S.S.; Cowart, M.; Benkovic, S.J. *Biochemistry* **1991**. *30*, 804–813.

48. Kuchta, R.D.; Benkovic, P.; Benkovic, S.J. *Biochemistry* **1988**, *27*, 6716–6725.

49. Petruska, J.; Sowers, L.C.; Goodman, M.F. *Proc. Natl. Acad. Sci. USA* **1986**, *83*, 1559–1562.

50. Fields, B.A.; Guss, J.M.; Freeman, H.C. *J. Mol. Biol.* **1991**, *222*, 1053–1065.

51. Greer, J. *Science* **1985**, *228*, 1055–1060.

52. Yeh, J.C.; Borchardt, R.T.; Vedani, A. *J. Comput.-Aided Mol. Design* **1991**, *5*, 213–234.

53. Argos, P.; Tucker, A.D.; Philipson, L. *Virology* **1986**, *149*, 208–216.

54. Steitz, T.A. Quart. Rev. *Biophysics* **1990**, *23*, 205–280.

55. Thomas, D.; Griffith, J. Furman, P.; Painter, G. *J. Cellular Biochem.* **1990**, *14D*, 112.

56. Darst, S.A.; Kubalek, E.W.; Kornberg, R.D. *Nature* **1989**, *340*, 730–732.

57. Painter, G.R.; Wright, L.L.; Andrews, C.W.; Cheng, N.; Hopkins, S.; Furman P.A. In *Advances in the Molecular Biology and Targeted Treatment for AIDS*, Kumar, A (ed.), Plenum, New York.

58. Merluzzi, V.J.; Hargrave, K.D.; Labadia, M.; Grozinger, K.; Skoog, M.; Wu, J.C.; Shih, C.; Eckner, K.; Hatlox, S.; Adams, J.; Rosenthal, A.S.; Taanes, R.; Eckner, R.J.; Koup, R.A.; Sullivan, J.L. *Science* **1990**, *250*, 1411–1413.

EPISULFONIUM IONS MAY NOT BE THE STEREODETERMINANTS IN GLYCOSYLATIONS OF 2-THIOALKYL PYRANOSIDES

Deborah K. Jones and Dennis Liotta

Advances in Molecular Modeling
Volume 3, pages 67–98.
Copyright © 1995 by JAI Press Inc.
All rights of reproduction in any form reserved.
ISBN: 1-55938-326-7

I. INTRODUCTION

Cyclic "onium" ions have been shown to form as the reactive intermediates in both (a) the electrophilic addition of certain reagents (e.g., Br_2,
Cl_2, HBr, HCl, $Hg(OAc)_2$, RSCl, RSeCl) to olefins, and (b) neighboring-
group participation reactions which occur during nucleophilic substitution reactions of certain substrates. The mechanistic pathways for (a) and
(b) are shown in Figure 1.

In the first step of pathway (a), electrophilic addition of the reagent
X–Y to the olefin double bond results in the formation of the intermediate
onium species. In the first step of pathway (b), the neighboring group X
acts as a nucleophile displacing the leaving group Z, while itself remaining attached to the molecule. The neighboring group can be either a lone
pair of nonbonding electrons from a single atom or from a group of atoms
such as C=O, C=C, aryl, and so forth. When X is participating in the

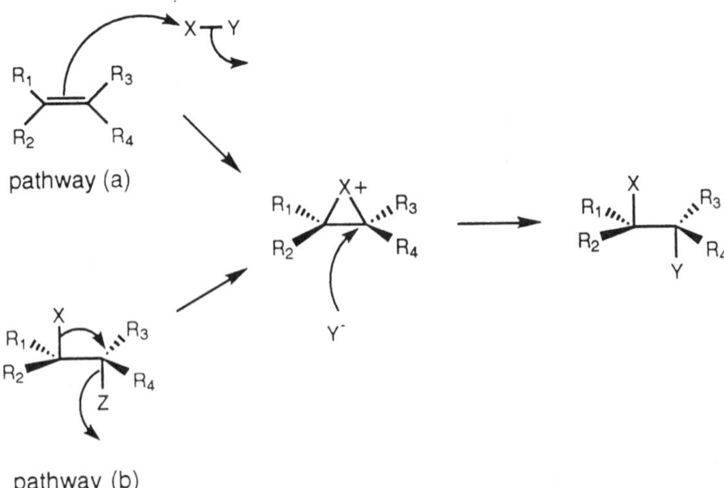

Figure 1. General mechanistic pathways involving "onium" ions.

bond-breaking stage, the neighboring group is said to be lending "anchimeric assistance." In the second step, orbital symmetry and steric considerations require that the external nucleophile Y^-, displaces X by a backside attack.

This type of positively-charged cyclic intermediate was first proposed in 1937 by Roberts and Kimball[1] to account for the stereochemistry observed in the halogenation of ethylenes [as represented by pathway (a) in Figure 1 where $XY = Br_2$, $X = Br^+$ and $Y = Br^-$]. The formation of an intermediate containing a halogen bridge (**2**, Figure 2) was postulated to rationalize why only *trans* products **5** were isolated. The previously

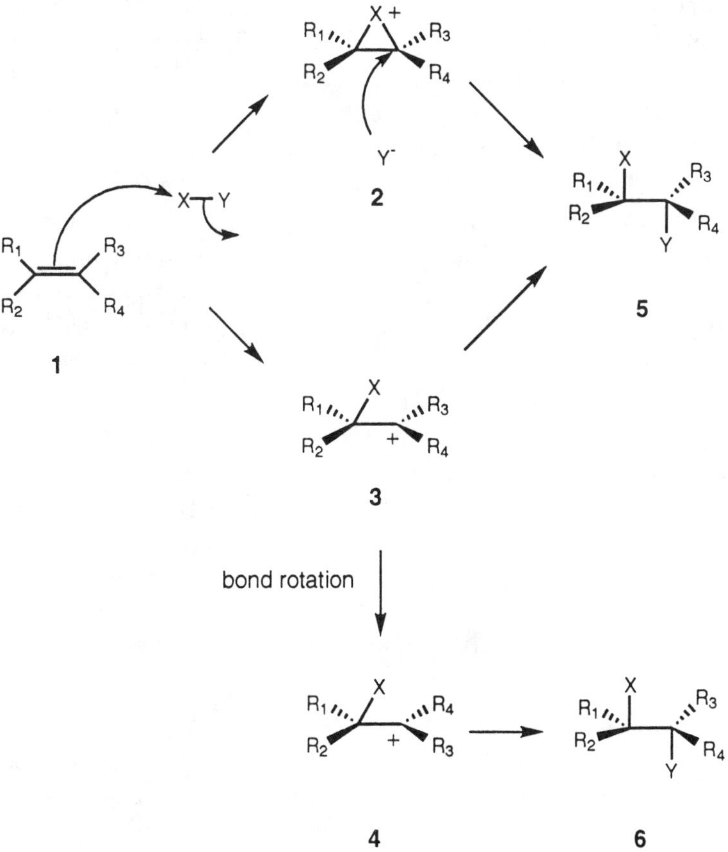

Figure 2. Rationale used by Roberts and Kimball.

suggested ion intermediate[2] **3** was discounted because of the presumption that free rotation about the carbon–carbon single bond from **3** to **4** would lead to a mixture of equal amounts of the *trans* and *cis* halogenation products, **5** and **6**, respectively.

This concept of "halonium" ion formation as the reactive intermediate was subsequently extended by Winstein and Lucas[3] to rationalize the outcome of S_N2 reactions occurring with overall retention of configuration [i.e., reactions occurring via the neighboring-group participation mechanism shown in pathway (b) in Figure 1]. They observed that the *threo* d,l pair of 3-bromo-2-butanol when treated with HBr gave dl-2,3-dibromobutane, while the *erythro* pair gave the *meso* isomer. In addition, either of the two *threo* isomers alone gave not just one of the enantiomeric bromides but the d,l pair. The rationale for this observation was that the intermediate bromonium ion present after the attack by the neighboring group was symmetrical, so that the external nucleophile Br^-, could equally well attack both carbon atoms.

The work by Winstein and Lucas[3] proved to be the starting point for a series of investigations into neighboring group effects in nucleophilic substitution reactions containing oxygen, nitrogen, sulfur, and halogens. Episulfonium ions (Figure 1, bridged species where X = SR or SH) were first proposed as chemical intermediates during studies into the neighboring group effect of sulfur in extremely reactive b-haloalkyl thioethers of the "mustard gas" type $S(CH_2CH_2Cl)_2$.[4]

Similarly in 1949, Kharasch and Buess[5] postulated a mechanism involving a cyclic episulfonium intermediate to explain the *trans* addition of 2,4-dinitrophenylsulphenyl chloride to olefins. Episulfonium ion intermediates have also been proposed to account for the unusual *anti*-Markovnikov orientation and *trans* stereochemistry of products from the addition of alkyl- and arylsulfenyl chlorides to olefins.[6] Both observations are consistent with a cyclic intermediate which is opened by chloride attack from a direction *trans* to the sulfur bridge.

Subsequent mechanistic and synthetic studies[7] have extensively investigated the nature of the episulfonium ion species for both cyclic and acyclic systems. Studies into the influence of substituents on the direction of episulfonium ion ring opening[7b–c] have shown that preferential attack occurs at the least substituted carbon. Alternatively, if the functional group (e.g., phenyl, vinyl, etc.) present at the more highly substi-

Figure 3. Possibility of oxonium ion formation.

tuted carbon center is able to effectively stabilize the developing partial positive charge, ring opening will occur at the more hindered position.

Investigations into the factors influencing episulfonium ion formation were carried out by measuring the relative rates of episulfonium ion formation.[7d–e] The observed reactivities reflect the opposing steric and electronic factors. For example, *trans*-2-butene is at least 24 times more reactive than *cis*-2-butene toward reaction with methanesulfenyl chloride.

For episulfonium ion intermediates in which there is an oxygen substituent adjacent to the three-membered sulfur containing ring, the possibility now exists for one of the oxygen lone pair of electrons to facilitate the opening of the episulfonium ion (Figure 3, **7**) to the oxonium ion **8**. Now the potential exists for addition of the nucleophile to the oxonium and/or the episulfonium ion.

Investigations into the addition of nucleophiles to α-sulfenyl acetals,[8a] monothioacetals and dithioacetals,[8b] and the electrophilic addition of benzenesulfenyl chloride to α,β-unsaturated ethers and sulfides[8c] examined this ring opening phenomenon for acyclic systems. Both the Okuyama[8a] and Otera[8b] groups have postulated that, depending on the reaction conditions and the nucleophile used, the conjugative stabilization (due to the adjacent alkoxy oxygen atom) of an open oxonium ion might make the contribution of an intermediate episulfonium ion species unnecessary.

The existence of episulfonium ions as reactive intermediates have also been postulated to rationalize the selective functionalization of cyclic oxygen-containing systems, specifically furanose and pyranoside derivatives.[9] For these cyclic systems the possibility of ring opening has in many cases not been addressed.

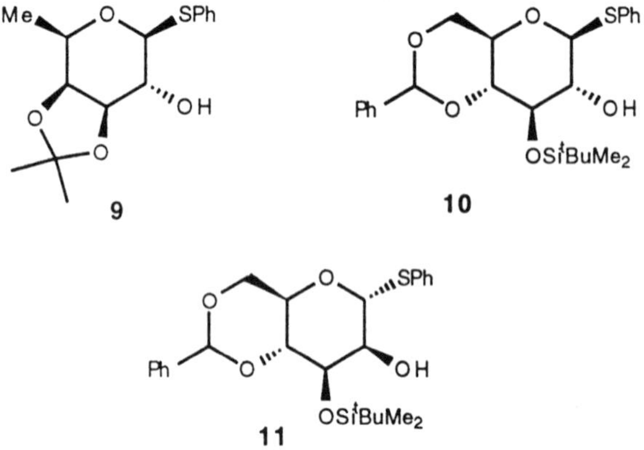

9 10 11

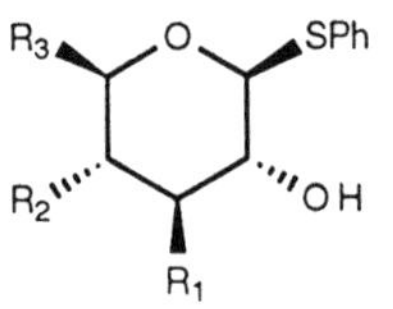

12 R_1 = OTBDMS, R_2 = OBn, R_3=CH_2OBn

13 R_1 = OTBDMS, R_2 = OMe, R_3 = CH_2OTBDPS

14 R_1 = H, R_2 =H, R_3 =H

Figure 4. Substrates investigated by Nicolaou and used for the basis of this computational study.

Nicolaou[9a] et al. investigated the stereospecific 1,2-migrations of the SPh group with a variety of pyranoside substrates (Figure 4, **9** to **13**). The fluorination of these substrates using diethylaminosulfur trifluoride (DAST) resulted in the exclusive formation of one glycosyl fluoride anomer. The stereoselectivity was rationalized by (a) intramolecular formation of an episulfonium ion and (b) a stereoelectronically-controlled ring opening by fluoride (see Figure 5).

Nicolaou did not consider the alternative mechanism (also shown in Figure 5) which involves the intramolecular ring-opening of the episulfonium ion, facilitated by the presence of the adjacent oxygen. In this mechanistic pathway, preferred addition of fluoride to one face of the oxonium ion due to the steric bias created by the substituents present on the glycosidic ring could also lead to the formation of one product.

The objective of this study was to use molecular orbital techniques to investigate the two alternative mechanistic pathways for the systems

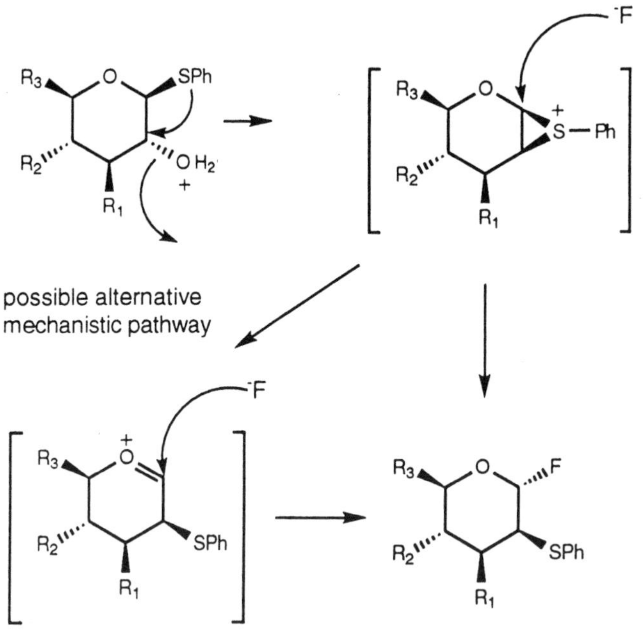

Figure 5. Nicolaou's rationale for the exclusive formation of one glycosyl fluoride anomer.

studied experimentally by Nicolaou et al. The general reaction sequence considered in evaluating the viability of these two pathways is shown in Figure 6.

Our approach was to initially locate the global minima for all starting materials, intermediates and products and subsequently locate all the relevant transition states for (a) episulfonium ion formation, (b) ring opening to the oxonium species, and (c) nucleophilic addition to both faces of the oxonium ion, with the SPh group either in an equatorial or an axial position to give the four possible fluorinated pyranoside anomers. The systems studied included the five molecules (Figure 4, **9** to **13**) investigated experimentally by Nicolaou et al.[9a] and an additional unsubstituted pyranose, (Figure 4, 14). The purpose of studying **14** was to have a unsubstituted model system by which to gauge the effects of the ring substituents on the selectivity of the reaction.

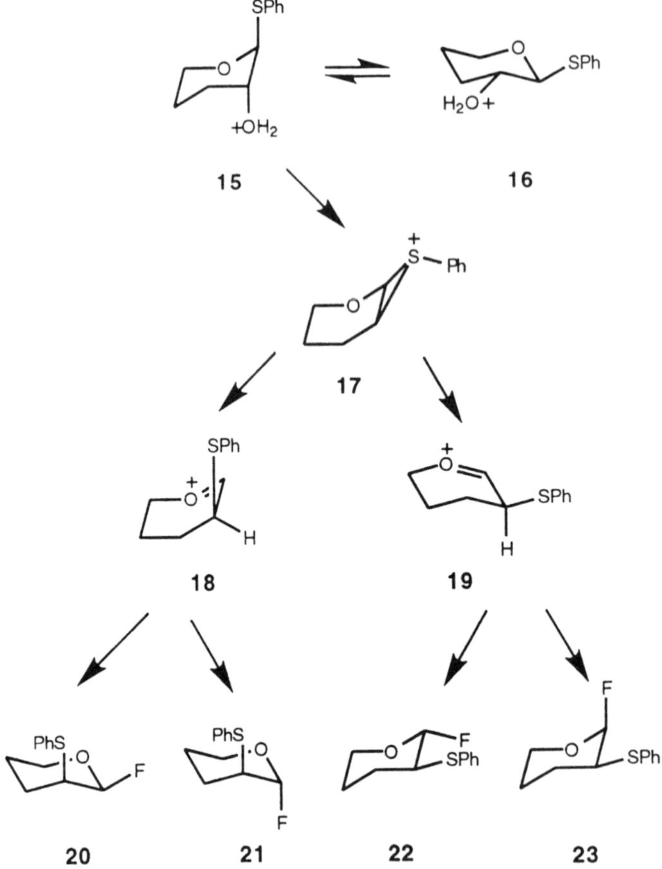

Figure 6. General reaction sequence evaluated computationally.

II. STARTING MATERIALS, EPISULFONIUM IONS, AND OXONIUM IONS

A. Methods

The input geometries for the minima to be used in the semiempirical calculations were obtained by optimizing each structure using the MMX force field.[10] This is the force field utilized in the molecular mechanics package PCMODEL 4.0.[11] The MMX force field is parameterized for use of the atom type S^+ in acyclic cases. Unfortunately, however, the appropriate bending and torsional parameters do not exist for atom type S^+ in a three-membered ring. Consequently, the episulfonium input geometry was approximated using an sp^3 nitrogen atom type. This atom type was edited to sulfur in the MOPAC input file, obtained from the geometry optimized using molecular mechanics. In addition, the $C–S^+$ bond lengths were altered to a more suitable starting value of 1.8 Å.

To eliminate rotamer problems and also to make the project more feasible in terms of CPU time, some of the protecting groups on the systems being studied were changed to smaller but suitably similar groups. For example, benzyl groups were calculated as ethyl groups and *t*butyldiphenylsilyl (TBDPS) groups as trimethylsilyl groups (TMS).

The subsequent molecular orbital calculations were carried out using MOPAC 5.0[12] and the MNDO Hamiltonian.[13] Since neither AM1 and PM3 are parameterized for sulfur, MNDO proved to be the most appropriate choice of Hamiltonian to use. The computations were performed on Silicon Graphics 4D / 35 and IBM 550 Work Stations.

The geometry optimizations in MOPAC on the starting materials, reaction intermediates, and products were done using the PRECISE option which increases the criteria for self-consistent field convergence by a factor of 100. This increases the default criteria for acceptance of a minimized geometry and, in general, gives more accurate results than calculations not employing PRECISE.

B. Results and Discussion

The resulting geometries were examined graphically using PCMODEL to ensure that no major geometric changes had occurred during optimization. Upon checking the episulfonium ion geometries, it

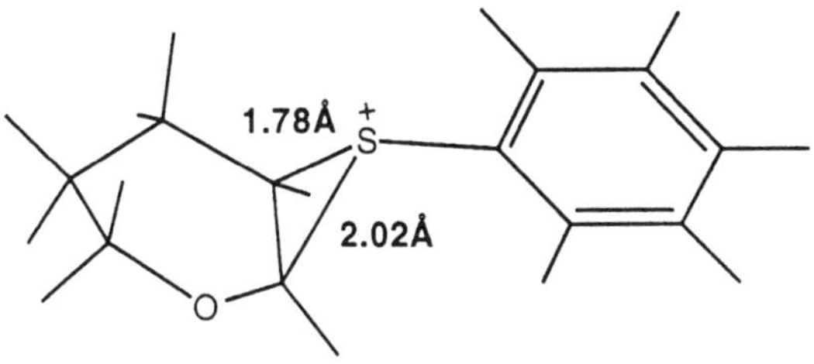

Figure 7. Optimized episulfonium ion geometry for model system **14**.

Table 1. MOPAC Results for Starting Materials and Oxonium Ions

	ΔH_f^0 *(kcal/mole)*			
	Diaxial[a] Starting Material	*Diequatorial[a] Starting Material*	*Diaxial Oxonium*	*Diequatorial Oxonium*
9	46.56	39.13	86.78	87.46
10	−23.03	−30.09	22.12	22.56
11	−33.68	−29.48	18.03	21.80
12	−76.52	−73.50	−32.63	−33.80
13	−138.85	−137.79	−96.30	−91.19
14	115.54	115.49	160.31	160.96

Note: [a]Starting material hydroxyl group calculated as being protonated.

was found that ring opening had occurred to give the oxonium ion. It was concluded that these type of episulfonium ions with an adjacent oxygen atom must in general lie in very shallow minima. The starting geometries generated from PCMODEL were sufficiently removed from the minima to cause optimization to the more stable oxonium ion species to occur.

In order to locate the episulfonium ion minima, reaction surfaces were generated around the general minimum area. This was done by systematically varying the two $C–S^+$ bond lengths from 1.75 Å to 2.10 Å in 0.05 Å increments. Episulfonium minimum location was confirmed by the results of a FORCE calculation. A system can be characterized as a ground state by the presence of six eigenvalues which are very small (less than about 30 cm^{-1}) present in the output from a FORCE calculation and all positive force constants. The episulfonium geometry for model system **14** is shown in Figure 7 (see Table 2 for other $C–S^+$ bond lengths).

The energy results of the MOPAC calculations are shown in Tables 1 and 2. In the case of the unsubstituted system **14**, the diaxial and diequatorial forms of the protonated starting material were approximately of the same stability (see Table 1). For acetonide **9**, the diaxial grouping forced the pyranose ring to adopt a boat conformation, which was energetically unfavorable compared to the diequatorial arrangement which adopts the corresponding chair conformation. The bicyclic systems **10** and **11**, displayed differing starting material conformation stability. The chair–chair conformation adopted by diequatorial **10** and diaxial **11** starting materials accounted for the greater stability observed

Table 2. MOPAC Results for Episulfonium Ions

	$\Delta H_f^{0\ a}$ Episulfonium	$\Delta H_f^{0\ a,b}$ (H_2O + Episulfonium)	C–S^+ Bond Length 1 (Å)	C–S^+ Bond Length 2 (Å)
9	94.71	40.41	1.99	1.79
10	23.71	–37.77	1.98	1.79
11	22.78	–38.16	1.99	1.79
12	–28.90	–89.60	2.00	1.79
13	–86.62	–147.56	2.00	1.79
14	166.72	105.78	2.02	1.78

Note: [a]Heat of formation in kcal/mole.

　　[b]ΔH_f^0 = –60.94 kcal/mole for water. This will allow direct comparison with the heat of formation for the protonated starting material.

(compared to the chair–boat conformation adopted by the corresponding diaxial **11** and diequatorial **12** starting materials). In the case of **12** and **13**, the relative stability of the two possible starting material conformations was not as in the previous examples governed by ring conformation, but by the bulky protecting groups being arranged in the most favorable conformation. Figure 8 compares the diaxial and diequatorial starting material conformations for **9, 10** and **12**.

Comparison of the heat of formation of either diaxial or diequatorial starting material with the heat of formation of the episulfonium plus water (see Tables 1 and 2) shows that in all cases, the products are more stable than the starting material. A similar comparison of just the episulfonium with either oxonium ion conformation shows in all cases that this ring opened product is more stable than the episulfonium ion. These results imply that with a favorable activation energy, both of these reactions (Figure 6, **15** to **17** and **17** to **18** or **19**) are viable.

III. FULLY CHARACTERIZING EPISULFONIUM AND OXONIUM ION FORMATION REACTIONS

A. Methods

The general approach taken for determining the activation energies for episulfonium ion and oxonium ion formation was to complete the full characterization of both reactions by locating the relevant transition

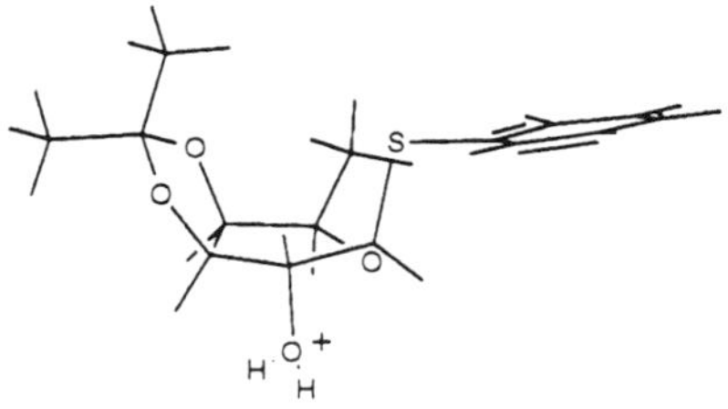

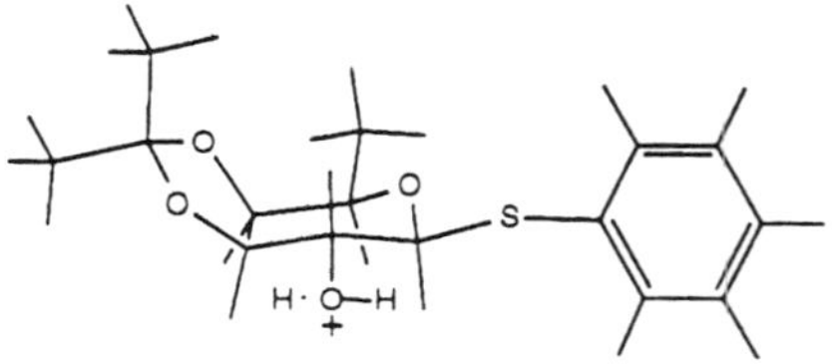

diequatorial 9
diaxial 9 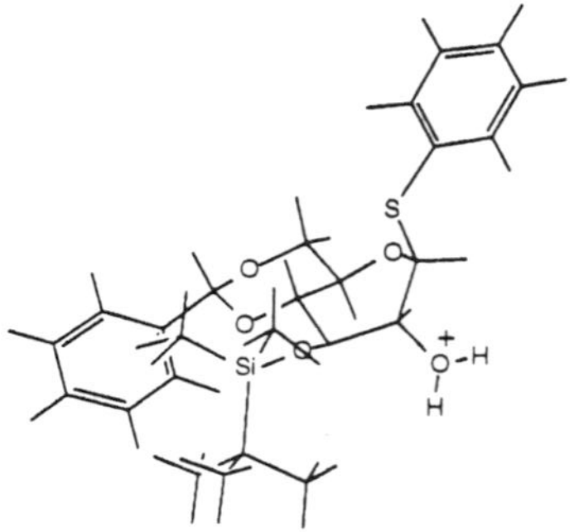
diequatorial 10
diaxial 10

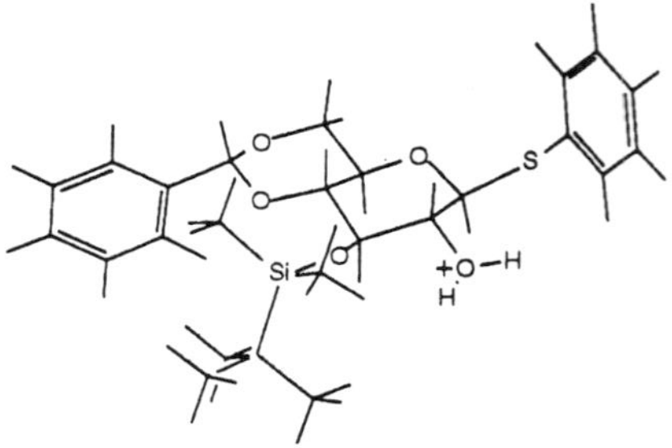

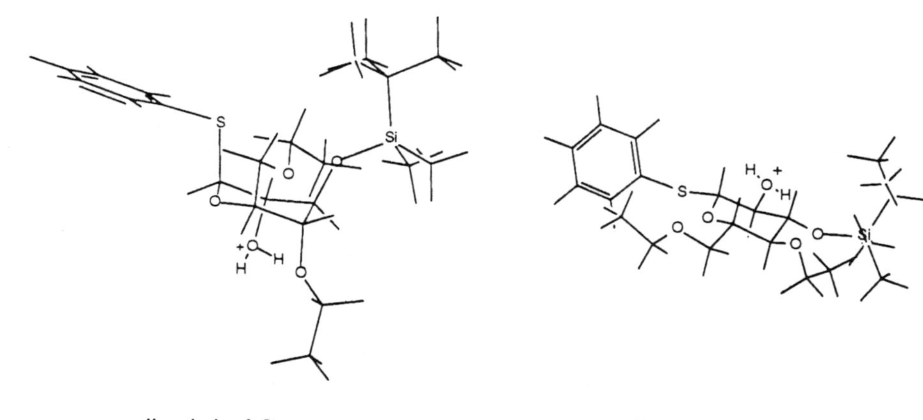

Figure 8. Comparison of the two possible starting material conformations.

states. Consider a general reaction written in the following way: A + B → [TS] → C + D, where A and B, the reactants, and C and D, the products, are stable species, and TS is the transition state located at the saddle point on the reaction coordinate connecting reactants to products. The first step in investigating the reaction is to fully optimize the geometries of the reactants and products and to confirm the location of minima. For the two reactions being considered this had already been completed. Once the geometries of reactants and products are defined, they can be used in the transition state location. If internal or Cartesian coordinates can be identified with the reaction coordinates, then by monotonically increasing or decreasing these coordinates, the reaction surface can be mapped with the transition state being the saddle point.

Since most molecular orbital methods can not explicitly handle solvation, serious errors can result when a substantial amount of charge is created in a reaction. To minimize this problem in the location of the episulfonium formation transition states, the starting material hydroxy groups were protonated, thereby permitting a conservation of charge going from starting material to products (Figure 6, **15** to **17**, episulfonium ion and water). The episulfonium ion formation reaction surface for **9** was generated by taking the diaxial starting material and varying C–S$^+$ bond 1 from 1.70 Å to 2.15 Å in 0.05 Å increments and C–S$^+$ bond 2 from 2.7 Å to 1.6 Å in 0.1 Å increments. The diaxial rather than the diequatorial conformation was used for the starting material since it had the correct geometry for the SN2 displacement which occurs in this reaction.

B. Results and Discussion

The reaction surface calculated (see Figure 9) was very steep and a discontinuity occurred across the whole surface when C–S$^+$ bond 2 = 2.1 Å. In general, such a discontinuity is an artifact of the computational method and the approach that is being taken to solve the problem. This normally indicates that another reaction coordinate, other than the two being varied to create the surface, is playing an important role in the reaction.

PCMODEL was used to graphically examine the geometries along the bond 1 reaction coordinate of 1.7 Å, at bond 2 values of 2.2, 2.1 and 2.0 Å. As bond 2 decreased in value, the C–O bond associated with loss of

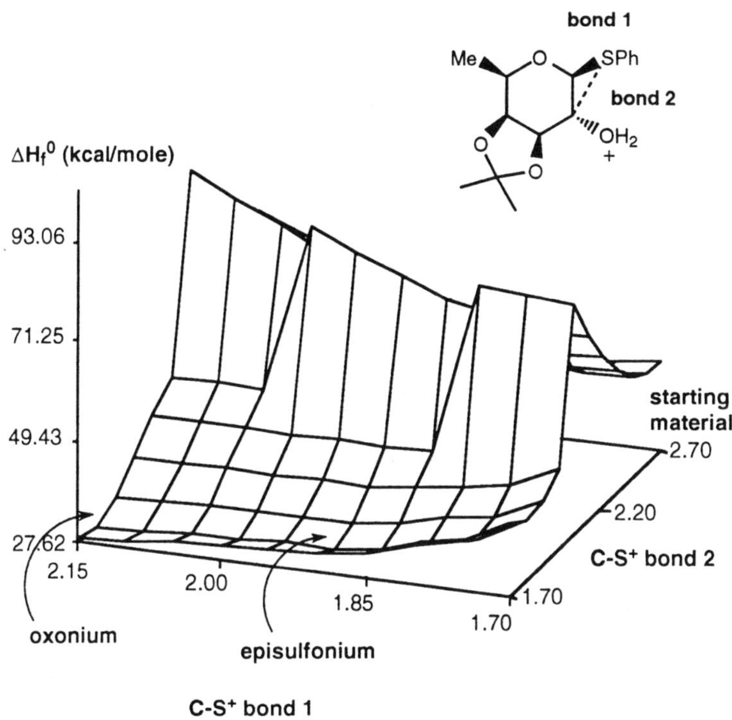

Figure 9. Discontinuous episulfonium formation surface for **9**.

a molecule of water increased greatly from 1.52 Å to 1.54 Å to 3.81 Å, respectively (i.e., spontaneous loss of water was responsible for the rapid decrease in energy and hence the discontinuity observed on the first reaction surface calculated, see Figure 10).

These results showed that the reaction coordinate associated with loss of the water molecule plays an important role in the characterization of the reaction surface. Since there were now three variables to be considered, the next approach taken to calculating a well-behaved reaction surface was to (1) fix C–S⁺ bond 1 with a value of 1.7 Å, since the first surface showed that it did not vary a great deal with any particular value of bond 2; (2) vary C–S⁺ bond 2 from 2.7 to 1.9 Å in 0.05 Å increments; (3) vary C–O bond length from 1.5 to 2.5 Å in 0.05 Å increments. Using this method, a saddle point was located on the surface produced (see Figure 11).

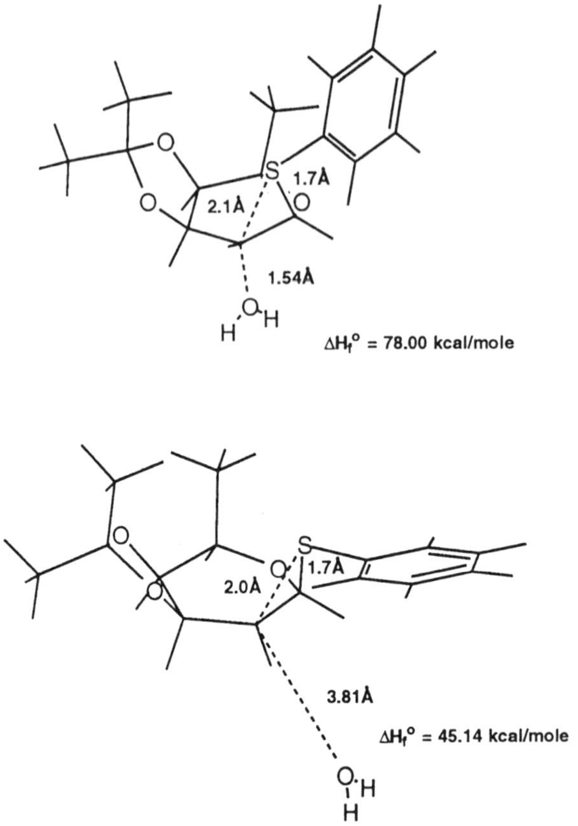

Figure 10. Loss of water in episulfonium ion formation for **9**.

Further refinement of the saddle point geometry obtained from the reaction surface was carried out by removing all the geometrical constraints and using the NLLSQ (nonlinear least squares) gradient minimization routine available in MOPAC. This keyword invokes minimization by Bartel's method, which is more suitable than the default minimization method for optimizing transition state geometries. Since only small changes occur during each minimization step using this method, this allows structural refinement, but prevents the delicate transition state geometry from collapsing to the more stable starting material or products. After obtaining a structure with a suitably low gradient norm

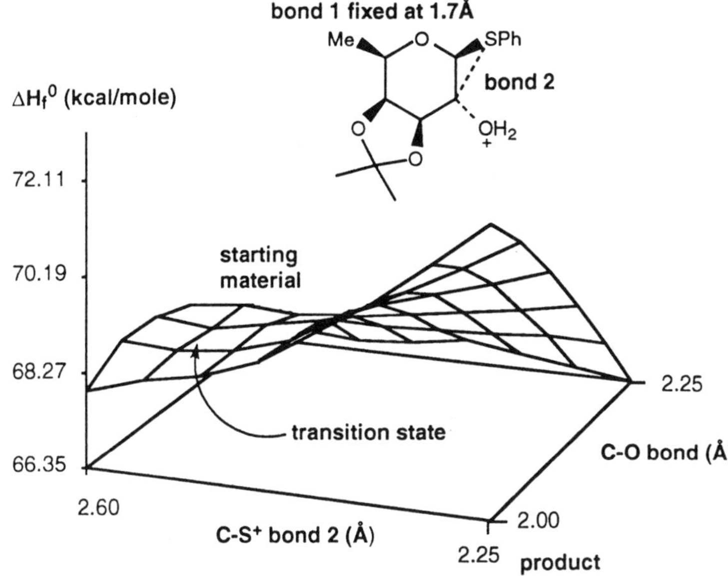

Figure 11. Reaction surface showing saddle point for **9**.

(GNORM), a FORCE calculation was carried out to evaluate all the force constants and confirm that a transition state had been located. A transition state must have exactly one negative force constant and the corresponding negative vibration must have components which correspond with the

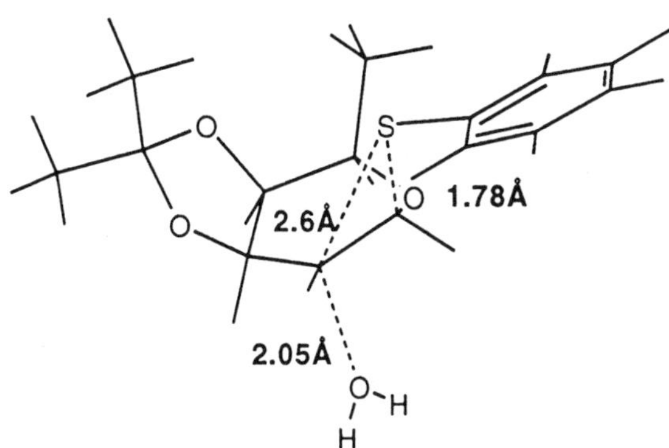

Figure 12. Episulfonium ion formation transition state located for **9**.

Table 3. Results for Formation of Episulfonium Ions and Water

	Bond 1 $C-S^+$ (Å)	Bond 2 $C-S^+$ (Å)	$C-O$ (Å)	ΔH_f^0 (TS) (kcal/mole)	$E_a{}^a$ (kcal/mole)
9	1.78	2.60	2.05	66.08	19.52
10	1.79	2.61	2.04	−5.55	17.48
11	1.78	2.61	2.11	−10.29	23.39
12	1.78	2.60	2.10	−54.55	21.94
13	1.78	2.58	2.11	−116.28	22.57
14	1.78	2.60	2.05	134.20	18.66

Note: $^a E_a = \Delta H_f^0$ (TS) − ΔH_f^0 (protonated diaxial starting material).

bonds being made and broken in the reaction being considered. The subsequent refined transition state geometry for **9** is shown in Figure 12.

This approach was used for locating the remaining episulfonium ion formation transition states for **10** to **14**. These results are shown in Table 3. The activation energy (E_a) varied from 17.48 to 23.39 kcal/mole and the three important reaction coordinates were similar in all the cases considered.

The activation energy (E_a) for ring opening of the episulfonium ion to the corresponding oxonium ion was determined by locating the transition state on the appropriate reaction surface. This surface was generated by systematically varying the two $C-S^+$ bonds comprising the three-membered episulfonium ion ring ($C-S^+$ bond 1 was varied from 1.75 Å to 2.25 Å in 0.05 Å increments and $C-S^+$ bond 2 from 2.00 Å to 1.75 Å in 0.05 Å increments).

Table 4. Results for Oxonium Ion Formation

	Bond 1 $C-S^+$ (Å)	Bond 2 $C-S^+$ (Å)	ΔH_f^0 (TS) (kcal/mole)	E_a (kcal/mole)
9	2.00	1.78	101.59	0.24
10	2.09	1.79	23.37	0.20
11	—	—	—	a
12	2.29	1.76	−28.46	0.44
13	—	—	—	a
14	—	—	—	a

Note: ^{a}Transition state not located in these examples.

The E_a for oxonium ion formation were all very small (see Table 4). For the examples in which it was possible to locate the transition state, the E_a values were of the same magnitude as the computational error normally associated with these calculations. These results suggest that episulfonium ion intermediate formation is followed by a facile ring-opening process to the more stable oxonium ion species.

IV. STUDY OF THE FLUORINATION OF OXONIUM IONS

A. Methods

The third part of this study involved investigating the addition of a nucleophile to both faces of the oxonium ions, with the SPh group either in an equatorial or an axial position. In the experimental work, DAST was used as the nucleophilic source of fluoride. It was not viable computationally to use DAST as the source of fluoride. Careful consideration was taken in choosing a computational surrogate for DAST.

Errors occur in semiempirical calculations when two charged species react to give a neutral species. Thus the simple choice of a fluoride ion as the nucleophile would have been problematic, since the negatively-charged fluoride ion and the positively-charged oxonium ions would readily collapse to product. By choosing hydrogen fluoride as the nucleophile, charge was maintained going from starting materials (oxonium ion plus HF) to products (fluorinated product plus H^+).

Initially considering example **9**, the starting point for location of the fluorination transition states was to optimize the four possible fluorinated products and confirm that minima had been found. The first reaction profile calculated was for the equatorial addition of HF to the oxonium with an equatorial SPh group. A proton was added to the diequatorial (SPh and F) product by editing the output geometry and this geometry was allowed to optimize so the proton adopted a favorable position. The key reaction coordinate in this process was the C–F bond length. To generate the reaction profile, the C–F bond in the product geometry (with the added proton) was varied from 1.35 Å to 2.00 Å in 0.05 Å increments.

The saddle point geometry was refined using the NLLSQ minimization technique to lower the GNORM. In addition to this, when the GNORM was below 5, the keyword SIGMA was used to complete the minimization, which invokes the McIver–Komornicki GNORM minimization routines, POWSQ and SEARCH. SIGMA first calculates a quite accurate Hessian matrix and then works out the direction of fastest descent and searches along that direction until the GNORM is minimized. The Hessian matrix is then partially updated in light of the new gradients and a fresh search direction is found. Unlike NLLSQ, utilizing this method allows GNORM minimization in both an up and down-hill direction in terms of energy.

The presence of one negative force constant in the results of the FORCE calculated confirmed the location of a transition state. This procedure was repeated to locate the three remaining possible transition states for **9**, and also for the other examples, **10** to **14**.

B. Results and Discussion

For **9** to **13**, the lowest energy fluorination transition state in all of these cases involved equatorial addition of hydrogen fluoride to the oxonium ion with an equatorial SPh substituent. This result was unexpected. We initially assumed that axial addition to the oxonium ion with an axial SPh substituent was going to be the lowest energy transition state, based on steric arguments. In the case of the model system **14**, the predicted axial addition to the axial SPh oxonium was indeed the lowest energy transition state, but by only 0.01 kcal/mole over the transition state with axial addition to the oxonium with an equatorial SPh substituent.

The fluorination transition state geometries were all very similar. The developing C–F bond varied by 0.12 Å between 1.66 and 1.78 Å, with the H–F bond of the fluoride source being constant at 0.97 Å in all of the transition states located. The angle of attack of the fluoride ion, given by the angle O–C–F, varied between 95.5° and 100.5°. This is consistent with the normal angle of approach in a nucleophilic addition reaction. The degree of pyramidalization of the carbon center to which the fluoride is adding is given by the dihedral $O-H-C_1-C_2$ (see Figure 13). Obviously, in the oxonium ion this dihedral had a value of almost 0° since the oxonium is planar and in the fluorinated product this dihedral had a value varying from approximately 32° to 38°. For the transition states located, the degree of pyramidalization was between 19.1° and 22.0°.

To assess the relative contribution of each transition state located to the overall product distribution, the vibrational, rotational and transla-

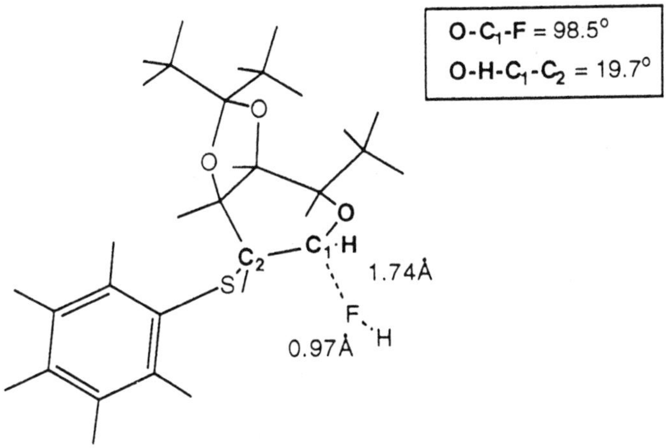

Figure 13.   Lowest energy fluorination transition state for **9**.

tional partition functions for all transition states and starting materials were computed. These thermodynamic properties are calculated during a FORCE calculation if the keywords THERMO(x) and ROT = y are specified (where x = the temperature at which these values are to be determined, i.e., the reaction temperature, and where y = the symmetry

Table 5.   Comparison of Theoretical and Experimental Product Distributions at 273 K

	k_{eqeq}[a] (b-pdt)	k_{eqax}[b] (a-pdt)	k_{axax}[c] (b-pdt)	k_{axeq}[d] (a-pdt)	*Exper.* b:a	*Theor.* b:a
9	1	4.06×10^2	2.90×10^4	2.00×10^6	100:0	96.4
10	1	1.05×10^6	4.12×10^1	2.33×10^7	100:0	100:0
11	1	3.70×10^3	3.00×10^4	1.30×10^5	100:0	99.6:0.4
12	1	4.00×10^2	1.67×10^2	2.00×10^5	100:0	96:4
13	1	1.69×10^3	2.45×10^5	2.35×10^6	100:0	99.8:0.2
14	5.04×10^1	6.02×10^1	1	2.30×10^3	[e]	71:29

Notes: [a] Equatorial addition of HF to the oxonium with an equatorial SPh.
[b] Axial addition of HF to the oxonium with an equatorial SPh.
[c] Axial addition of HF to the oxonium with an axial SPh.
[d] Equatorial addition of HF to the oxonium with an equatorial SPh.
[e] Experimental data not available for this compound.

Table 6. Thermodynamic Properties of **9**

	$\Delta H_f^{0\ a}$	$E_a^{\ a}$	Partition Functions[c]	
			$Q_{vib} \times 10^6$	$Q_{rot} \times 10^6$
Starting Materials[b]				
eq. SPh oxonium + HF	27.73	—	3.786	5.11
ax. SPh oxonium + HF	27.05	—	24.130	5.04
Transition States				
eq. SPh, eq. addition of HF (major product)	44.57	16.84	13.800	6.01
eq. SPh, ax. addition of HF (minor product)	46.15	18.42	10.720	5.78
ax. SPh, ax. addition of HF (major product)	48.27	21.22	8.457	5.79
ax. SPh, eq. addition of HF (minor product)	52.10	25.05	6.840	5.57

Notes: [a]Units of kcal/mole.
[b]For hydrogen fluoride: $\Delta H_f^0 = -59.73$ kcal/mole and partition functions, $Q_{vib} = 1.0$ and $Q_{rot} = 4.93$.
[c]Partition functions evaluated at 0 °C.

Table 7. Thermodynamic Properties of **10**

	$\Delta H_f^{0\ a}$	$E_a^{\ a}$	Partition Functions[c]	
			$Q_{vib} \times 10^6$	$Q_{rot} \times 10^6$
Starting Materials[b]				
eq. SPh oxonium + HF	-37.17	—	3.631	2.75
ax. SPh oxonium + HF	-37.61	—	1.071	2.52
Transition States				
eq. SPh, eq. addition of HF (major product)	-19.52	17.65	14.910	2.95
eq. SPh, ax. addition of HF (minor product)	-12.28	25.33	24.240	2.68
ax. SPh, ax. addition of HF (major product)	-19.41	18.20	4.961	2.72
ax. SPh, eq. addition of HF (minor product)	-11.57	26.24	7.810	2.68

Notes: [a]Units of kcal/mole.
[b]For hydrogen fluoride: $\Delta H_f^0 = -59.73$ kcal/mole and partition functions, $Q_{vib} = 1.0$ and $Q_{rot} = 4.93$.
[c]Partition functions evaluated at 0 °C.

Table 8. Thermodynamic Properties of **11**

	$\Delta H_f^{0\,a}$	$E_a{}^a$	Partition Functions^c	
			$Q_{vib} \times 10^6$	$Q_{rot} \times 10^6$
Starting Materials^b				
eq. SPh oxonium + HF	–37.93	—	5.785	2.38
ax. SPh oxonium + HF	–41.70	—	15.050	2.70
Transition States				
eq. SPh, eq. addition of HF (major product)	–22.96	14.97	33.810	2.87
eq. SPh, ax. addition of HF (minor product)	–20.86	17.07	6.477	2.68
ax. SPh, ax. addition of HF (major product)	–22.96	18.74	33.810	2.87
ax. SPh, eq. addition of HF (minor product)	–19.80	21.90	48.010	2.82

Notes:　^{a}Units of kcal/mole.
^{b}For hydrogen fluoride: $\Delta H_f^0 = -59.73$ kcal/mole and partition functions, $Q_{vib} = 1.0$ and $Q_{rot} = 4.93$.
^{c}Partition functions evaluated at 0 °C.

Table 9. Thermodynamic Properties of **12**

	$\Delta H_f^{0\,a}$	$E_a{}^a$	Partition Functions^c	
			$Q_{vib} \times 10^6$	$Q_{rot} \times 10^6$
Starting Materials^b				
eq. SPh oxonium + HF	–93.53	—	4.055	2.02
ax. SPh oxonium + HF	–92.36	—	1.492	1.80
Transition States				
eq. SPh, eq. addition of HF (major product)	–74.11	19.42	21.770	2.18
eq. SPh, ax. addition of HF (minor product)	–73.23	20.30	4.582	2.10
ax. SPh, ax. addition of HF (major product)	–71.24	21.12	3.130	1.91
ax. SPh, eq. addition of HF (minor product)	–67.63	24.73	2.693	2.03

Notes:　^{a}Units of kcal/mole.
^{b}For hydrogen fluoride: $\Delta H_f^0 = -59.73$ kcal/mole and partition functions, $Q_{vib} = 1.0$ and $Q_{rot} = 4.93$.
^{c}Partition functions evaluated at 0 °C.

Table 10. Thermodynamic Properties of **13**

	$\Delta H_f^{0\ a}$	$E_a{}^a$	Partition Functions[c]	
			$Q_{vib} \times 10^6$	$Q_{rot} \times 10^6$
Starting Materials[b]				
eq. SPh oxonium + HF	−150.92	—	2.749	3.01
ax. SPh oxonium + HF	−156.03	—	7.501	2.63
Transition States				
eq. SPh, eq. addition of HF (major product)	−136.17	14.75	3.249	2.64
eq. SPh, ax. addition of HF (minor product)	−132.80	18.12	2.732	2.64
ax. SPh, ax. addition of HF (major product)	−135.85	20.18	4.203	2.65
ax. SPh, eq. addition of HF (minor product)	−133.01	23.02	68.770	2.91

Notes: [a]Units of kcal/mole.

[b]For hydrogen fluoride: $\Delta H_f^0 = -59.73$ kcal/mole and partition functions, $Q_{vib} = 1.0$ and $Q_{rot} = 4.93$.

[c]Partition functions evaluated at 0 °C.

Table 11. Thermodynamic Properties of **14**

	$\Delta H_f^{0\ a}$	$E_a{}^a$	Partition Functions[c]	
			$Q_{vib} \times 10^6$	$Q_{rot} \times 10^6$
Starting Materials[b]				
eq. SPh oxonium + HF	−101.23	—	1.669	1.66
ax. SPh oxonium + HF	−100.58	—	0.816	1.59
Transition States				
eq. SPh, eq. addition of HF (major product)	44.57	18.43	13.580	2.24
eq. SPh, ax. addition of HF (minor product)	46.15	17.67	4.009	2.23
ax. SPh, ax. addition of HF (major product)	48.27	17.66	3.165	2.16
ax. SPh, eq. addition of HF (minor product)	52.10	21.39	7.250	2.10

Notes: [a]Units of kcal/mole.

[b]For hydrogen fluoride: $\Delta H_f^0 = -59.73$ kcal/mole and partition functions, $Q_{vib} = 1.0$ and $Q_{rot} = 4.93$.

[c]Partition functions evaluated at 0 °C.

number of the molecule, i.e., the number of equivalent positions attainable by pure rotations). The results of these calculations and the relevant heats of formation and subsequent activation energies are shown in Tables 5 to 10.

Transition state theory was used to analyze the partition functions (see Tables 6 to 11) and hence determine the relative contribution of each transition state.[15] The normalized rate constants, k, for each pathway and experiment $\beta{:}\alpha$ ratios are shown in Table 5. The calculated $\beta{:}\alpha$ product ratios were in excellent agreement with the experimental results obtained by the Nicolaou group (100% β-product).

V. AB INITIO STUDY OF MODEL SYSTEMS

A. Methods

Up to this point, the semiempirical results suggested that a mechanism invoking oxonium ions as the reactive intermediates in glycosylations of 2-thioalkyl pyranosides had been shown to be a viable alternative to the postulated episulfonium intermediates. To add further evidence for this postulate, ab initio calculations using GAUSSIAN 92[16] were performed.

GAUSSIAN 92 offers a choice of basis sets. The basis sets used are all of the class called Gaussian Type Orbital (GTO). These orbitals use Gaussian functions to approximate the individual atomic orbitals, which are then used in constructing the linear combination of atomic orbitals (LCAO) to form the MOs. The use of GTOs is an approximation which improves as the number of GTOs used per atomic orbital increases. These GTOs are also referred to as basis functions. By increasing the number of basis functions, the accuracy of the calculation is improved, but CPU time and disk space increase roughly as N^4 where N is the total number of basis functions used.

The simplest of the available basis sets in GAUSSIAN 92 are called STO-nG (Slater Type Orbitals approximated by n Gaussians). The smallest basis set commonly used is the STO-3G. The STO-nG basis sets are called minimal basis sets because they provide the minimum number of basis functions necessary to accommodate all the electrons and maintain overall spherical symmetry. The advantage of using a minimal basis set is the savings in CPU time and disk space requirements as each hydrogen has a single basis function (corresponding to the 1s orbital) and the first row atoms use five basis functions (1s, 2s, 2px, 2py, 2pz). The disadvantage of this basis set is its inability to adapt to different environments about an atom.

The next more sophisticated basis set is the split-valence basis set. In this basis set, the atomic orbitals corresponding to the valence shell are split such that there are two basis functions per valence shell orbital where each of the basis function coefficients can be optimized in the LCAO procedure. These basis sets are typically written as n-21G or n-31G where n is the number of Gaussian functions used for the core orbitals (counted as a single function per core orbital) and the 21 and 31 describe how the valence shell orbitals are split (2 Gaussians + 1 Gaussian or 3 Gaussians + 1 Gaussian). These are counted as 2 basis functions per valence shell orbital.

The addition of d orbitals for all nonhydrogen atoms gives the next improvement in the accuracy of the basis set. These orbitals are called polarization functions because they are not normally involved directly in bonding but rather allow the center of the p orbitals to shift away from the atomic center. An example of such a basis set is the 6-31G*. This basis set uses a 6 Gaussian contraction for the core orbitals with a 3/1 split for the valence shell orbitals (both s and p). The asterisk indicates that a single set of six d orbitals are also included. An extension of this polarization basis set is the 6-31G** in which a set of p functions have been added to the hydrogen basis functions. Polarization functions are important for describing small ring systems and molecules containing the second row elements.

The only ab initio study of episulfonium ions[17] in the literature, is by Csizmadia et al. Using the STO-3G and 4-31G basis sets, calculations were carried out on a cyclic sulfurane **26** and a thiiranium-chloride ion-pair **25** (see Figure 14). Their objective was to compare what they considered to be the two possible mechanisms for sulphenyl halide addition to olefins. Using these basis sets the episulfonium ion **25** was characterized as a minimum.

The objective of these ab initio calculations was to locate the episulfonium ion and oxonium ion for the unsubstituted tetrahydropyran with a SMe substituent, to confirm their existence as minima (see **28** and **29**, Figure 15) and then to compare the relative stability of the these two intermediate species. Choosing **28** and **29** as the molecules to be studied

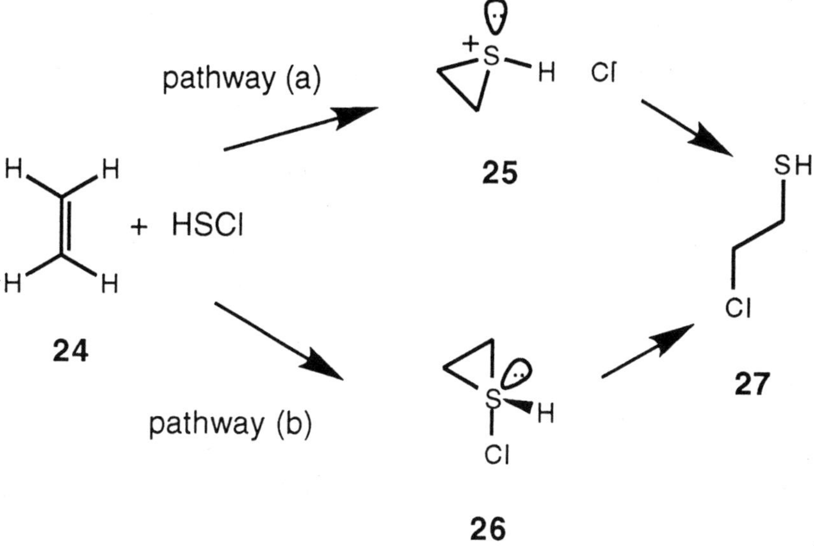

Figure 14. Systems studied by Csizmadia et al.

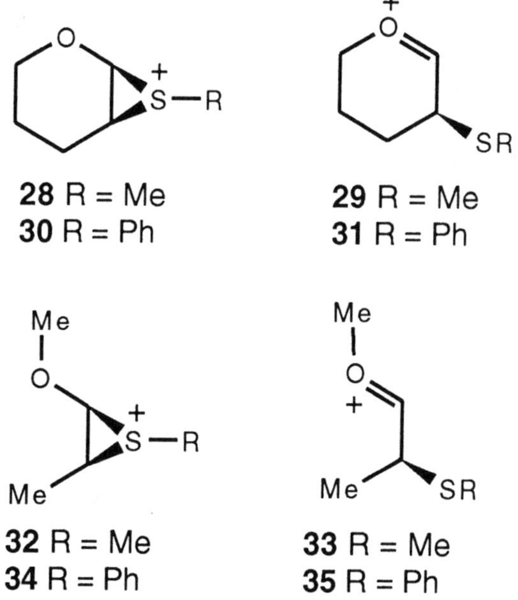

Figure 15. Molecules studied using GAUSSIAN 92.

kept the number of atoms small enough (8 heavy and 11 light atoms) to make such a CPU intensive calculation feasible.

To generate input geometries for the ab initio calculations on **33** and **34**, MOPAC calculations were performed. This proved to be problematic since the episulfonium ion **28** was not located as a minimum. However upon optimization it opened to the oxonium ion form **29**. Consequently, the MNDO minimized geometries from **14** (SPh substituent) were used as the input structure for the ab initio calculation of **28**. The phenyl group was substituted by a methyl group using PCMODEL.

The Hartree–Fock (HF) type calculations were carried out using both STO-3G and 6-31G* basis sets. The STO-3G basis set was chosen to initially give quick results, to be followed by, and compared with, the results of the lengthier 6-31G* basis set calculations which take into account the d-orbitals in the molecule. The structures **28** and **29** were optimized, and subsequent FORCE calculations confirmed the location of minima.

B. Results and Discussion

Examination of the optimized geometries from the STO-3G calculations using PCMODEL showed that the episulfonium ion structure **28**, had been located as a minimum. Interestingly, the oxonium ions **29**, with SMe in the axial or equatorial positions, optimized to the corresponding episulfonium ion structure (see Table 12). Conversely, the optimized geometries from the 6-31G* calculations showed that the episulfonium ion optimized to the corresponding oxonium ion, but both equatorial and axial oxonium ions were located as minima. Comparing the results of oxonium ion geometry optimization using 6-31G* to MNDO showed that the structures were very similar.

The calculations which had optimized to the corresponding structures were repeated using the following procedures: (a) initially fixing the two C–S bonds in the structures and then removing the geometrical constraints and allowing the structures to fully optimize; (b) carrying out the calculations using Cartesian coordinates instead of the Z-matrix format which had been previously used; (c) fully calculating the force constants associated with the molecule at each step of the optimization, instead of using analytic gradients which is the default procedure. All of these methods had no effect on the final outcome of these calculations.

Table 12. Results for GAUSSIAN 92 Calculations

	MNDO (kcal/mole)	*STO-3G (a.u)*	*6-31G* (a.u)*
28	[a]	–697.7155952	[a]
29 (ax. SMe)	131.49	[b]	–705.7309367
29 (eq. SMe)	131.99	[b]	–705.7291814
30	166.72	–885.8914296	[a]
31 (eq. SPh)	160.31	[b]	[c]
31 (eq. SMe)	160.96	[b]	[c]
32	[a]	–660.2806301	[a]
33	134.40	[b]	–667.851417
34	166.31	–848.4569362	[a]
35	163.14	[b]	[c]

Notes: [a]Episulfonium ion optimizes to the corresponding oxonium ion.

[b]Oxonium ion optimizes to the corresponding episulfonium ion.

[c]After extended periods of CPU time, geometries did not achieve the optimization criteria.

The possibility existed that by changing the sulfur substituent from the original phenyl group, which was successfully used in the semiempirical calculations, to a methyl group to save CPU time, had somehow perturbed the episulfonium ion and oxonium ion geometries. Repeating the two sets of ab initio calculations on phenyl substituted **30** and **31** also resulted in the same phenomena as observed for **28** and **29** (see Table 12). Calculations on both methyl and phenyl substituted acyclic systems, **32** and **33**, and, **34** and **35**, respectively, in an attempt to remove a strain inherent in the previous cyclic systems calculated, again resulted in only episulfonium ions being located using the STO-3G basis set and oxonium ions being located using the 6-31G* basis set.

VI. CONCLUSIONS

The semiempirical results using MOPAC 5.0 and the MNDO Hamiltonian suggest that the 1,2-migration of the SPh group may be occurring via formation of a transient episulfonium ion species, but this undergoes rapid opening to form the more stable oxonium ion species (see Figure 16).

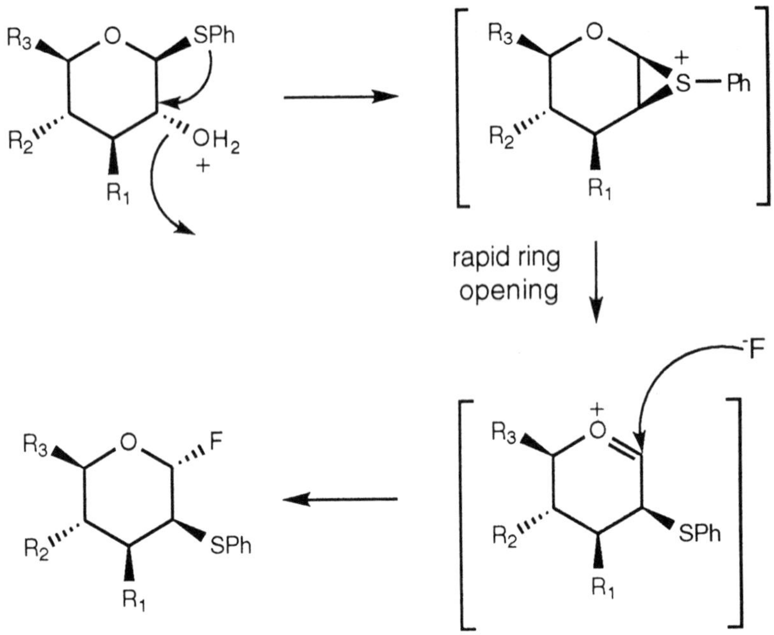

Figure 16. Overall mechanism in agreement with the computational results.

The excellent agreement between the experimental and these computational results suggests that oxonium ions, not episulfonium ions, are the actual stereodeterminants in this reaction manifold. Interestingly, a comparison of the calculated $\beta{:}\alpha$ ratio of the simple model system **14**, with any of the experimentally-employed substrates, **9** to **13**, suggests that only modest selectivity is predicted for the unsubstituted molecule. Thus, according to our model, the observed selectivity is a consequence of conformational restrictions imposed by ring substituents, rather than stereoelectronic factors inherent in nucleophilic openings of episulfonium ions.

In conclusion, a mechanism invoking oxonium ions as the reactive intermediates in glycosylations of 2-thioalkyl pyranosides has been shown by semiempirical methods to be a viable alternative to the postulated episulfonium intermediates.

The results of the ab initio calculations proved to be inconclusive. Using the simpler STO-3G basis set, for any system with an oxygen adjacent to the episulfonium ion ring system, episulfonium ion forms were located but the corresponding open oxonium ion forms were never located as minima. Conversely, using the more complex 6-31G* basis set, which is thought to give a better description of small cyclic systems and for systems containing second row elements, oxonium ions were located as minima, but the corresponding episulfonium ions were never located. These results held true, regardless of whether the substituent on the sulfur atom was a methyl or phenyl group, or the system being studied was cyclic or acyclic.

Without having located an episulfonium ion and the corresponding oxonium ion using one basis set, it was not possible to draw any conclusions as to the relative stability of the two forms from these higher level quantum-mechanical calculations. However, one interpretation of the 6-31G* results which is consistent with the semiempirical results is that the episulfonium ion "well" is so shallow that it simply can not be located. This again suggests that oxonium ions are the stereodeterminants.

REFERENCES

1. Roberts, I.; Kimball, G.E. *J. Am. Chem. Soc.* **1937**, *59*, 947.
2. Ingold. *Chem. Rev.* **1934**, *15*, 225.
3. (a) Winstein, S.; Lucas, H.J. *J. Am. Chem. Soc.* **1939**, *61*, 1576. (b) Winstein, S.; Lucas, H.J. *J. Am. Chem. Soc.* **1939**, *61*, 2845.
4. (a) Fuson, R.C.; Price, C.C.; Burness, D.M. *J. Org. Chem.* **1946**, *11*, 475. (b) Bartlett, P.D.; Swain, C.G. *J. Am. Chem. Soc.* **1949**, *71*, 1406.
5. Kharasch, N.; Buess, C.M. *J. Am. Chem. Soc.* **1949**, *71*, 2724.
6. Mueller, W.H.; Butler, P.E. *J. Am. Chem. Soc.* **1966**, *88*, 2866.
7. (a) Senning, A.; Lawesson, S.O. *Tetrahedron* **1963**, *19*, 695. (b) Thaler, W.A.; Mueller, W.A.; Butler, P.E. *J. Am. Chem. Soc.* **1968**, *90*, 2069. (c) Mueller, W.H.; Butler, P.E. *J. Am. Chem. Soc.* **1968**, *90*, 2075. (d) Brown, C.; Hogg, D.R. *J. Chem. Soc., Chem. Commun.* **1965**, 357. (e) Thaler, W.A. *J. Org. Chem.* **1969**, *34*, 871. (f) Mueller, W.H. *Angew. Chem. Int. Ed.* **1969**, *81*, 482. (g) Smit, W.A.; Zefirov, N.S.; Bodrikov, I.V.; Krimer, M.Z. *Acc. Chem. Res.* **1979**, *12*, 282. (h) Kamimura, A.; Sasatani, H.; Hashimoto, T.; Ono, N. *J. Org. Chem.* **1989**, *54*, 4998. (i) Hogg, D.R. *Mech. React. Sulfur Compounds* **1970**, *5*, 87. (j) Lucchini, V.; Modena, G.; Pasquato, L. *J. Am. Chem. Soc.* **1991**, *113*, 6600.
8. (a) Saigo, K.; Kudo, K.; Hashimoto, Y.; Kimoto, H.; Hasegawa, M. *Chem. Lett.* **1990**, 941. (b) Sato, T.; Otera, J.; Nozaki, H. *J. Org. Chem.* **1990**, *55*, 6116. (c) Toyoshima, K.; Okuyama, T.; Fueno, T. *J. Org. Chem.* **1978**, *43*, 2789.

9.   (a) Nicolaou, K.C.; Ladduwahetty, T.; Randall, J.L.; Chucholowski, A. *J. Am. Chem. Soc.* **1986**, *108*, 2466. (b) White, J.D.; Theramongkol, P.; Kuroda, C.; Engebrecht, J.R. *J. Org. Chem.* **1988**, *53*, 5909. (c) Ramesh, S.; Kaila, N.; Grewal, G.; Franck, R.W. *J. Org. Chem.* **1990**, *55*, 5. (d) Nicolaou, K.C.; Hummel, C.W.; Bockovich, N.J.; Wong, C-H. *J. Chem. Soc., Chem. Commun.* **1991**, 870. (e) Ogawa, T.; Ito, Y. *Tetrahedron Lett.* **1987**, *28*, 2723. (f) Pruess, R.; Schmidt, R.R. *Synthesis* **1988**, 694. (g) Baldwin, M.J.; Brown, R.K. *Can. J. Chem.* **1968**, *46*, 1093. (h) Baldwin, M.J.; Brown, R.K. *Can. J. Chem.* **1967**, *45*, 1195. (i) Smoliakova, I.P.; Smit, W.A.; Osinov, B. *Tetrahedron Lett.* **1991**, *32*, 2601. (j) Ramesh, S.; Franck, R.W. *J. Chem. Soc., Chem. Commun.* **1989**, 960. (k) Ryan, K.J.; Acton, E.M.; Goodman, L. *J. Org. Chem.* **1971**, *36*, 2646. (l) Kondo, T.; Abe, H.; Goto, T. *Chem. Lett.* **1988**, 1657. (m) Ogawa, T.; Ito, Y. *Tetrahedron Lett.* **1988**, *29*, 3987. (n) Zuurmond, H.M.; van der Klein, P.A.M.; van der Marel, G.A.; van Boom, J.H. *Tetrahedron Lett.* **1992**, *33*, 2063. (o) Ito, Y.; Numata, M.; Sugimoto, M.; Ogawa, T. *J. Am. Chem. Soc.* **1989**, *111*, 8508.

10. Gilbert, K.; Gajewski, R. In *Advances in Molecular Modeling*, Vol. 1, D. Liotta (ed.), JAI Press, Greenwich, CT, 1990.

11. We are grateful to Serena Software for making this software available to us.

12. (a) Stewart, J.J.P. *J. Comp. Aided. Des.* 1990, *4*, 1. QCPE Program No. 455; (b) Jones, D.K.; Liotta, D.C. *Tetrahedron Lett.* **1993**, *34*, 7209.

13. (a) Dewar, M.J.S.; Thiel, W. *J. Am. Chem. Soc.* **1977**, *99*, 4899. (b) Dewar, M.J.S.; Rzepa, H.S. *J. Am. Chem. Soc.* **1978**, *100*, 777. (c) Dewar, M.J.S.; Reynolds, C.H. *J. Comp. Chem.* **1986**, *7*, 140. (d) Dewar, M.J.S.; Friedheim, J.; Grady, G.; Healy, E.F.; Stewart, J.J.P. *Organometallics* **1986**, *5*, 375.

14. Gucker, F.T.; Seifert, R.L *Physical Chemistry*, W. W. Norton, New York.

15. A detailed theoretical explanation of the approach used to calculate the relative contribution of each transition state to the overall product distribution has been given in Chapter 1 concerning oxazaborolidines.

16. Gaussian 92, Revision B: M.J. Frisch, G.W. Trucks, M. Head-Gordon, P.M.W. Gill, M.W. Wong, J.B. Foresman, B.G. Johnson, H.B. Schlegel, M.A. Robb, E.S. Replogle, R. Gomperts, J.L. Andres, K. Raghavachari, J.S. Binkley, J.J.P Stewart, and J.A. Pople, Gaussian, Inc., Pittsburgh PA, 1992.

17. Csizmadia, V.M.; Schmid, G.H.; Mezey, P.G.; Csizmadia, I.G. *J. C. S. Perkin Trans II* **1977**, 1019.

STEREOSPECIFIC SYNTHESIS OF 1β-METHYLCARBAPENEM INTERMEDIATES:

UNDERSTANDING THE SELECTIVITY OBSERVED

Deborah K. Jones and Dennis Liotta

Advances in Molecular Modeling
Volume 3, pages 99–144.
Copyright © 1995 by JAI Press Inc.
All rights of reproduction in any form reserved.
ISBN: 1-55938-326-7

I. INTRODUCTION

The discovery of thienamycin (Figure 1, **1**), a carbapenem antibiotic, promoted extensive research into this class of β-lactam compounds. The two major deficiencies associated with thienamycin are: (1) its chemical instability at high concentration and (2) its susceptibility to renal dehydropeptidase-I (DHP-I).

In an attempt to overcome these problems, the semi-synthetic structural modification of thienamycin by Shih et al. at Merck yielded a chemically stable product, imipenem (Figure 1, **2**), which had the antibacterial activity and potency of the parent compound. Unfortunately, it also was readily metabolized by DHP-I, making it necessary to co-administer a DHP-I inhibitor with this antibiotic. None of the subsequent thienamycin analogs have improved biological and chemical stability while retaining good antibacterial activity.

In further efforts to improve the biological and physical properties of these antibiotics, the Merck group[1] synthesized the 1β-methyl-substituted carbapenem, (–)-(1*R*, 5*S*,6*S*)-2-(2-*N*,*N*-dimethylamino-2-iminoethyl-thio)-6-[(1*R*)-1-hydroxyethyl]-1-methylcarbapen-2-em-3-carbox

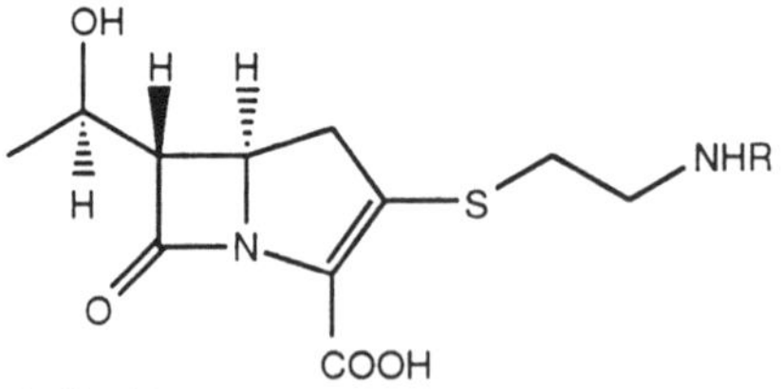

1, R = H
2, R = CH=NH

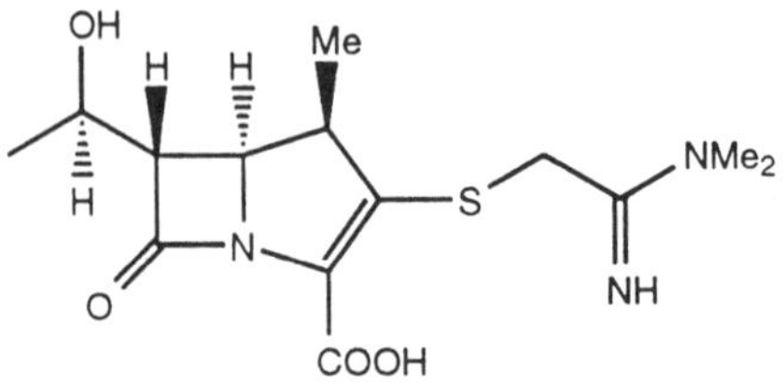

3, First 1β-methyl carbapenem synthesized

Figure 1. Thienamycin and related analogs.

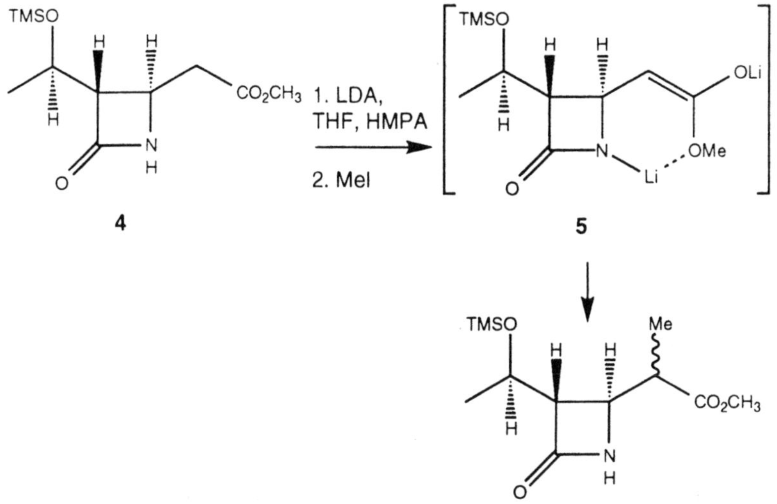

Figure 2. First synthesis of a 1-Me-substituted carbapenem.

ylic acid (Figure 1, **3**). The Merck group's primary objective was to obtain the 1α- and 1β-methyl compounds for structure activity studies. The synthetic route used to introduce the methyl group, gave a ratio of 4:1, α-Me:β-Me (see Figure 2) and subsequent biological testing of **3** showed that the introduction of the 1β-methyl moiety improved the biological activity of this antibiotic.

After showing the importance of the 1β-methyl moiety, the Merck group synthesized both 1α-methyl- and 1β-methylthienamycin. Biological studies showed that 1β-methylthienamycin retained the antibacterial activity of thienamycin and was highly resistant to hydrolysis by DHP-I; the 1α-methyl isomer was resistant to DHP-I hydrolysis, but its antibacterial activities were very much decreased.[2] These results validated the effect of incorporating the methyl substituent and also the importance of the relative stereochemistry of this group.

Many stereoselective syntheses of the key 1β-methyl intermediate **6**, have subsequently been reported.[3-14] The first synthesis to follow the work of the Merck group was by a Japanese group,[3] the key synthetic step being the reaction of acetoxyazetidinone **7**, with ketene silyl acetal

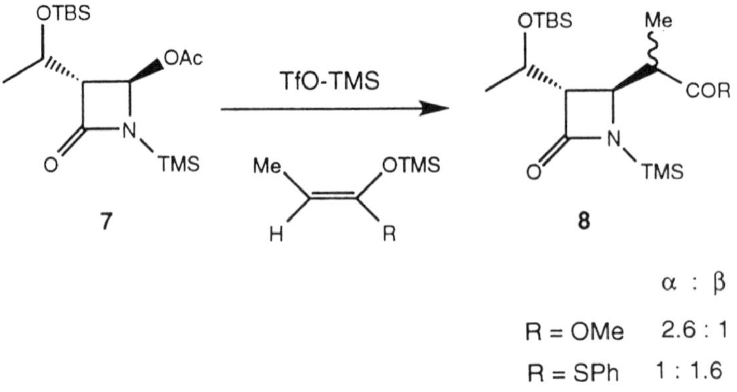

Figure 3. Synthesis used by Shibata et al.

in the presence of trimethylsilyltrifluoromethane sulfonate (Figure 3). In common with the method utilized by Shih et al., this synthesis did not result in a good yield of the desired β-isomer.

Many of the subsequent methods used for the stereoselective synthesis of the key intermediate **6**, have involved the use of reagents that are either expensive or difficult to handle on a large scale. For example, the approach adopted by Noyori[5] to the synthesis of the methylated precursor **10**, involved using the BINAP-Ru(III) catalyzed asymmetric hydrogenation which gave a quantitative yield and a β:α ratio of 99.9:0.1 (Figure 4). Unfortunately, although this method gives excellent selectivity, economically it is not suitable for use industrially due to the high price of both the Ru(III)-catalyst and BINAP ligand.

The same disadvantage applies to the strategy employed by the Merck group[8] which uses chiral oxazolidinone enolates. This highly successful double-asymmetric synthesis lead to the 1β-methyl carbapenem precur-

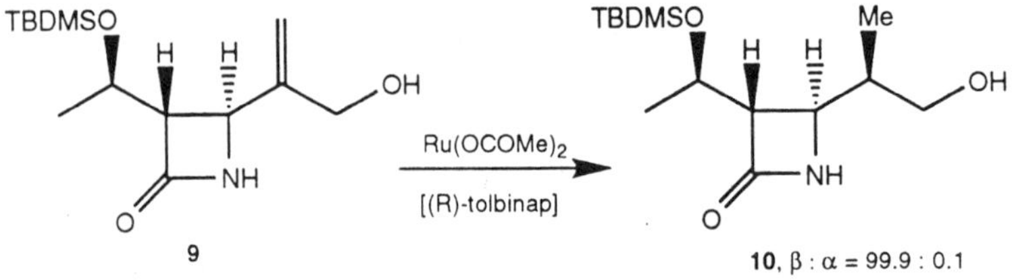

Figure 4. Asymmetric hydrogenation employed by Noyori et al.

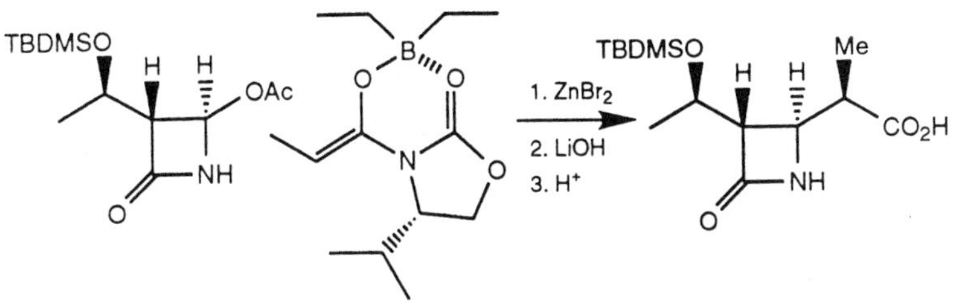

Figure 5. Use of chiral oxazolidinone enolates.

sor **13**, in >98% enantiomeric excess (Figure 5). The fact that the chiral auxiliary would have to be removed and recycled makes this approach not viable for use on a large scale industrial process, despite the high selectivity observed.

Further studies by the Merck group[9] into the stereocontrolled chiral synthesis of 1β-methylcarbapenem precursors, resulted in the development of catalytic hydrogenation conditions to introduce the 1β-methyl moiety. Using Raney Ni as the catalyst, a ratio of 95:5 was observed for the hydrogenation of the exocyclic double bond in the 1-azabicyclo[4.2.0]octane systems investigated (see Figure 6, **14**). The major disadvantage of this method, even though the β:α ratio is good, relates to the availability of the 1-azabicyclo[4.2.0]octane starting materials. By far the most favorable starting material for the synthesis of 1β-methyl carbapenem antibiotics, on an industrial scale, is 3-siloxyethyl-4-acetoxyazetidinone (**11**, Figure 5). This starting material is commercially available and very inexpensive.

Recently, Bender et al. reported[15] a highly diastereoselective method for epimerization of α-methyl ester **16a** to the corresponding β-methyl isomer **16b**, via enolborate formation and subsequent kinetic protonation from the less hindered α-face of the enolate **17** (Figure 7).

In search of a simple, scalable and highly diastereoselective synthesis of the key methylated intermediate **16b**, Choi et al. envisioned that a similar transformation[16] might also be possible by decarboxylation of a diacid (**18**, Figure 8), followed by stereoselective protonation of the resultant ketene acetal **19**. The structural similarity seen between the two

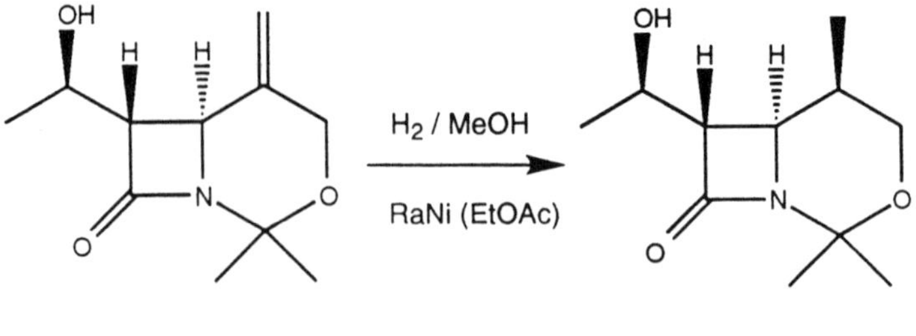

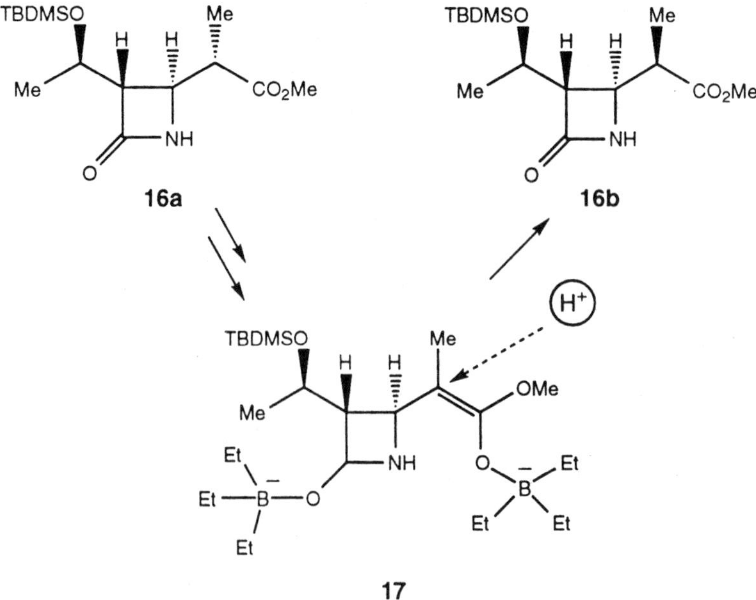

Figure 6. Catalytic hydrogenation.

intermediates, enolate **17** and ketene acetal **19**, suggests that similar kinetic protonation stereochemistries would be likely.

The dicarboxylic acid **23a**, was prepared from commercially available starting material, 3-silyloxyethyl-4-acetoxyazetidinone **20** and 2,2,5-trimethyl-1,3-dioxan-4,6-dione **22** (Figure 9), giving **22a** in 95% yield.

Figure 7. Diastereoselective epimerization by Bender et al.

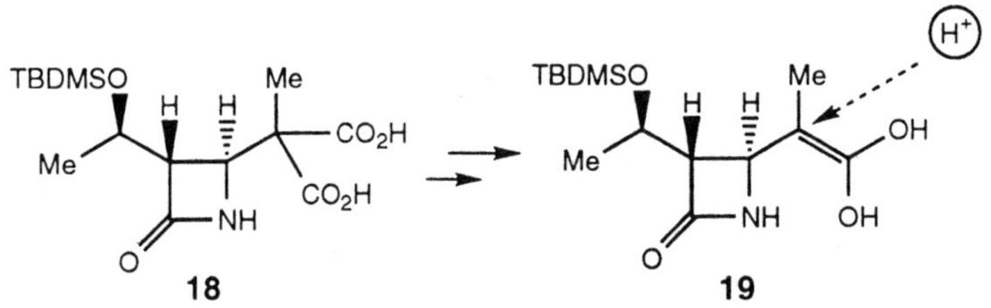

Figure 8. Rationale behind the possible diastereoselective decarboxylation.

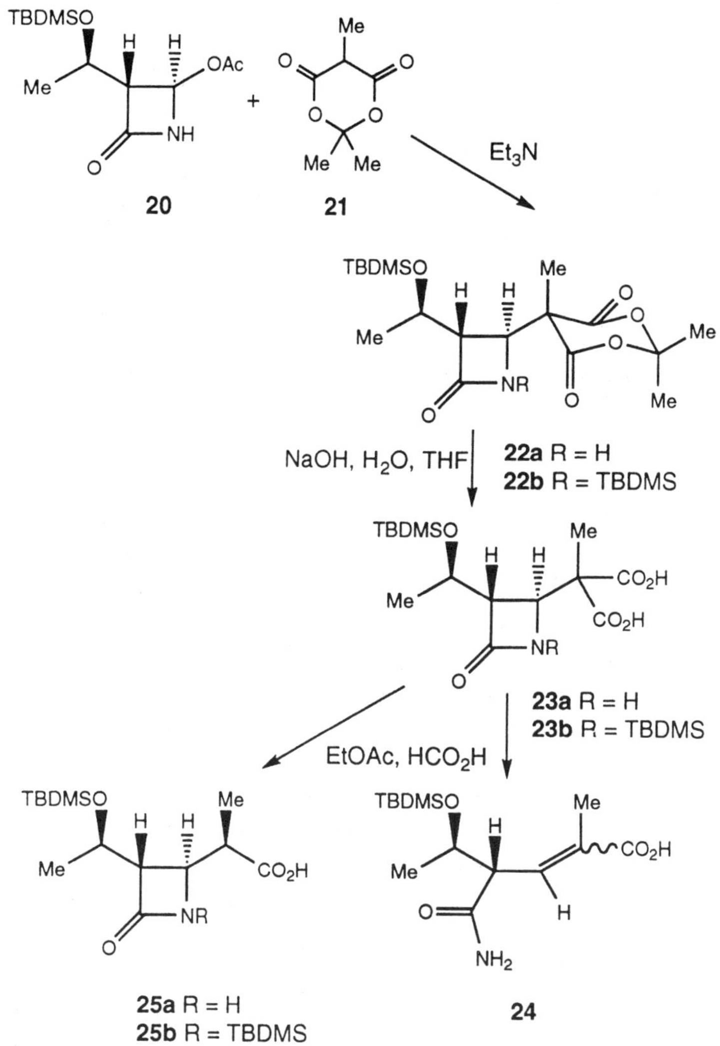

Figure 9. Synthesis of the diacid and subsequent decarboxylation.

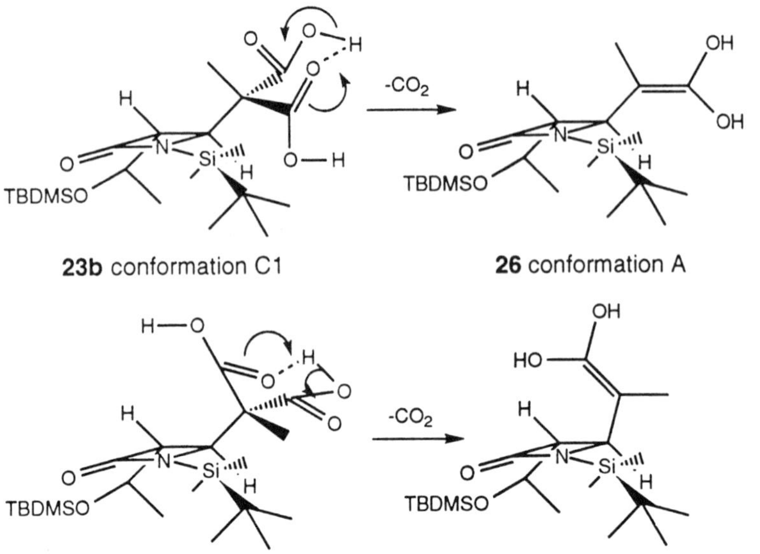

23b conformation C1

26 conformation A

23b conformation D1

26 conformation B

Figure 10. Rationale for protonation selectivity.

The diacid was obtained by ring opening of **22a**. However, when the diacid **23a**, was subjected to the decarboxylation conditions, the major product was the ring opened amide **24**. To suppress this process, the lactam nitrogen was silylated. Base hydrolysis of **22b** gave the *N*-silylated diacid **23b**. When diacid **23b** was subjected to the same decarboxylation conditions, the reaction proceeded in a highly stereoselective manner, providing **25b** as a 94:6 ratio of β:α isomers. Removal of the protecting group and recrystallization gave **25a** in 64% yield, >99% β-product.

The observed protonation selectivity was rationalized by the Merck group as follows. Assuming that the ketene acetal **26b** could be formed by decarboxylation of either carboxyl group of diacid **23b**, the bulky *N*-silyl group of ketene acetal **26b** forces the ketene acetal to orient in one of two conformations, A or B (Figure 10). Conformation A is expected to be preferred due to the unfavorable A-1,3 strain between the ketene acetal hydroxy group and the C_3 methine proton in conformation B. Protonation of A would be expected to occur from the least hindered α-face (the β-face is blocked by the bulky *N*-silyl group) resulting in the formation of the desired β-methyl acid **25a**. In the absence of the large *N*-silyl group, it is assumed that the ketene acetal can more easily adopt a conformation in which the C_4–N bond in the lactam ring is orthogonal to the ketene acetal π-bond resulting in elimination to give amide **24**.

Further experiments were carried out, in which addition of methoxide to the Meldrum's acid adduct **22b**, followed by acidification gave a diastereomeric mixture of two methyl ester acids (5:1). Curiously, only one of these diastereomers underwent decarboxylation to give the methyl ester of **25a** (β:α = 94:6); the other diastereomer remained unreacted even after 3 days at 80°C. This suggested the possibility that the reactive conformation of the ketene acetal **26b**, is the result of a diastereospecific decarboxylation of diacid **23b**. The decarboxylation may be controlled by the conformation of the precursor diacid **23b** and the law of microscopic reversibility, that is, just as protonation of **26b** can only occur from the less-hindered α-face, loss of CO_2 likewise can only occur from the same face, away from the *N*-silyl group. The diacid of conformation C1

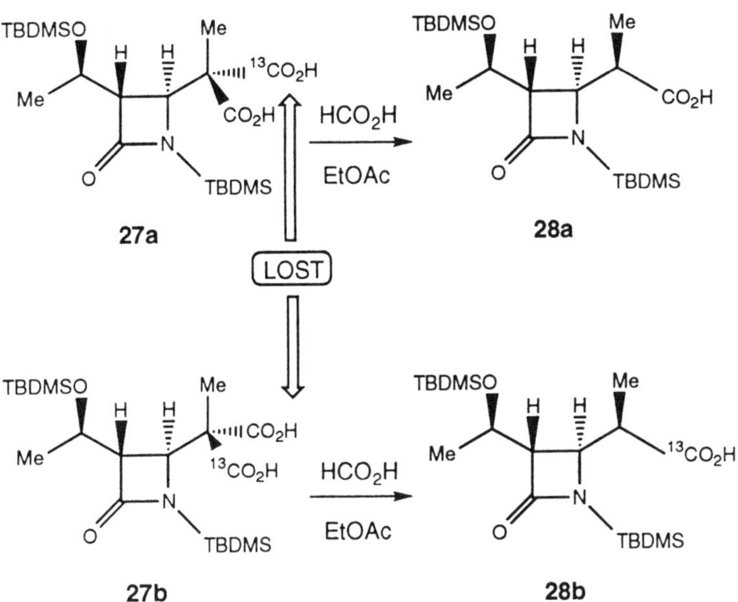

Figure 11. ^{13}C-labeling studies.

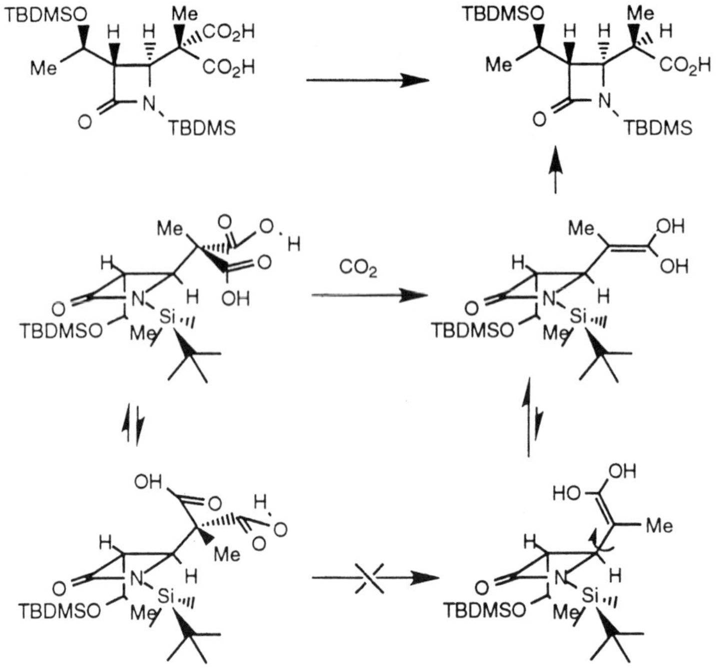

Figure 12. Proposed mechanism.

undergoes a diastereospecific decarboxylation of the pro (*S*) carboxyl to give ketene acetal **26b** in the preferred A conformation, which would then be protonated from the least hindered α-face to give the desired β-methyl acid **25b**. Conversely, the diacid conformation D1 would be expected to undergo diastereospecific decarboxylation of the pro (R) carboxyl to give the undesired α-methyl acid **25b** via selective protonation of the less stable ketene acetal conformation B.

To gain more insight into this proposed diastereospecific decarboxylation, ^{13}C labeling studies of the diacid were carried out (see Figure 11). The two ^{13}C labeled isomers, **27a** and **27b**, were prepared and subjected to the decarboxylation conditions.

It was found that one isomer completely lost the labeled carboxylic acid group giving **28a**, while the other isomer completely retained the labeled carboxylic acid group giving **28b**. This result suggested that only one of the two carboxylic acid groups was lost, that is, a stereospecific decarboxylation of the diacid was occurring.

The overall proposed mechanism taking into account both a stereospecific decarboxylation and a stereospecific protonation is shown in Figure 12.

The objective of this computational study[19b] was to evaluate the mechanism proposed by the Merck group (see Figure 12) using semi-empirical methods. The investigation was comprised of two distinct parts: study of the loss of CO_2 to form the ketene acetal and its protonation to form the β-methyl acid. The results and discussion are presented in this order also.

II. METHODS

For both the decarboxylation and protonation studies, the calculations were carried out in three stages: (1) the study of a model system, **29a** and **29b**; (2) the study of a simplified β-lactam system, **30a** and **30b**; (3) and the study of the system used in the experimental work, **23b** and **26** (see Figure 13).

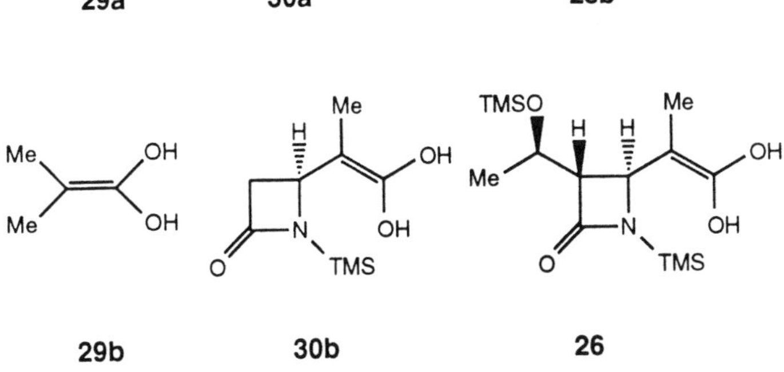

Figure 13. Molecules used in the decarboxylation and protonation studies.

Use of a model system allowed the computational approaches to be tested and refined on a molecule with a small number of atoms, which kept the CPU time to a minimum. Application of the results from the model system **29** to **30** allowed a preliminary mechanistic investigation to be carried out. (The assumption was made that the side chain was removed from the site of reactivity in the molecule, thus could be neglected). Finally, study of **23b** and **26** checked the validity of studying these systems without the β-lactam side chain. In all calculations, *t*-butyldimethylsilyl (TBDMS) protecting groups were replaced by trimethylsilyl (TMS) groups, to reduce CPU time and to eliminate unnecessary rotamer problems.

The structures investigated in this study were constructed in PCMODEL[17] and minimized using the MMX force field[18] available in this molecular mechanics package. These minimized structures were used as the input geometries for the subsequent semi-empirical molecular orbital calculations.

For the calculations involving **23b** and **26**, which both have the conformationally mobile side-chain, the GRIDSEARCH routine available in SYBYL 6.0 was used. GRIDSEARCH performs a systematic search on all bonds selected, and minimizes each conformation, holding the torsion angles of the bonds being searched fixed. The number of conformations generated increases with the number of bonds, so GRID-SEARCH is limited in practicality to searching only a few bonds. Only two torsion angles were searched (see Figure 13, which defines these two torsions) at 30° increments which generated only 144 possible conformers. The lowest energy conformer located was used as the input geometry for the subsequent molecular orbital calculations.

When this procedure was applied to potential transition state structures, the key bond lengths, angles and torsions intimately involved in the transition state (e.g., the C–H bond length of the proton being added to the ketene acetal in the protonation studies) were fixed using the CONSTRAINT option. This allowed the GRIDSEARCH to search the conformational space for the side chain and do the appropriate minimizations, without disrupting the "transition state part" of the molecule.

Theoretical treatment of these systems was done with MOPAC 5.0[19] using the AM1 Hamiltonian.[20] The computations were performed on Silicon Graphics Personal Iris 4D/35 and IBM 550 workstations. Included in each calculation were the keywords, PRECISE and GNORM = 0.1 to insure both good geometrical and thermodynamical data.

When optimizing transition state geometries, the NLLSQ and SIGMA keywords were employed as more suitable minimization techniques. Potential transition state geometries that had two negative force constants instead of the requisite one negative force constant required for a transition state were studied using the intrinsic reaction coordinate (IRC) method pioneered and developed by Gordon.[21] This technique is used to annihilate stray vibrations which result in undesired second or sometimes third negative force constants. The IRC defines the path the atoms in a molecule would follow if the system was allowed to relax, starting from the transition state geometry. The resultant forces on each atom are used, together with the isotopic mass of each atom, to calculate which direction to move the atoms, in order to obtain a lower energy state. A normal coordinate is chosen, corresponding to the vibrations associated with the negative force constant to be eliminated, thereby allowing location of the desired transition state.

The sequence of operations to start an IRC calculation is as follows: (1) do a normal FORCE calculation on the transition state with the stray negative force constants, specifying ISOTOPE in order to save the FORCE matrices and allow subsequent calculations to be restarted without having to recalculate the FORCE matrices; (2) using the keywords IRC = n and RESTART run the IRC calculation, the value of n being selected to correspond to the normal mode which needs to be eliminated and the sign associated with n being either positive or negative depending on the direction of displacement (i.e., either the positive or negative direction along this coordinate); (3) optimization of the new geometry resulting from the IRC calculation using SIGMA or NLLSQ.

III. RESULTS AND DISCUSSION

A. Decarboxylation Studies

Model System 29a

The mechanism proposed for loss of CO_2 from the diacid **29a** resulting in ketene acetal formation, **29b**, was via a six-membered transition state (see Figure 14). An appropriate choice of reaction coordinates to be

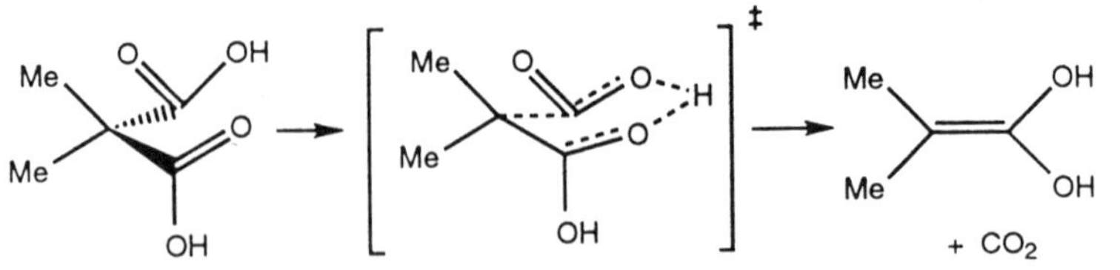

29a   **29b**

$\Delta H_f^0 = -194.27$ kcal/mole $\Delta H_f^0 = -180.25$ kcal/mole

Figure 14. Proposed six-membered decarboxylation transition state.

varied in the generation of the decarboxylation reaction surface was crucial, and with six bonds varying in this transition state, the choice was made even more complex.

A survey of the literature revealed that the majority of decarboxylation computational studies were carried out at the ab initio level and studied small mono-acid molecules such as formic acid,[22] acetic acid,[23] pyruvic acid,[24] and glyoxylic acid.[25] Ab initio studies on the decarboxylation of the 1,2-diacid, oxalic acid[26] considered a mechanistic pathway of intramolecular proton transfer followed by loss of CO_2. This was not a viable mechanistic pathway for the 1,3-diacid **29a**.

Results of a mass spectroscopy study of polyether antibiotics containing a β-hemiketal carboxylic acid group showed that an unusual fragmentation process occurred where both CO_2 and H_2O were lost from the molecular ion. Subsequent ab initio studies on model systems of these antibiotics investigated possible mechanistic routes. One of the pathways studied was the loss of CO_2 and H_2O via a six-membered transition state (see Figure 15). The obvious similarities to the transition state of interest

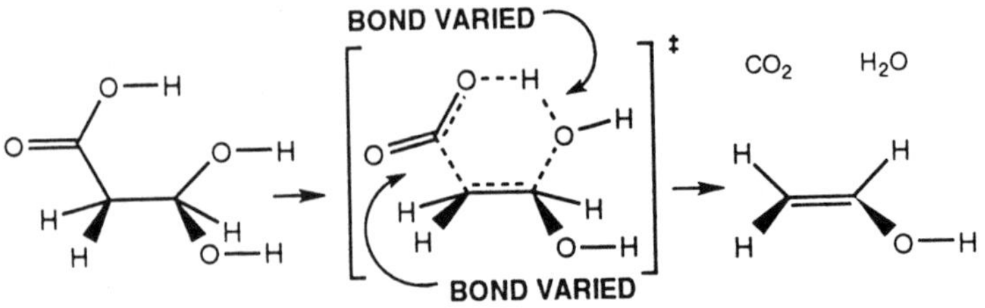

Figure 15. Transition state studied by Siegel et al.

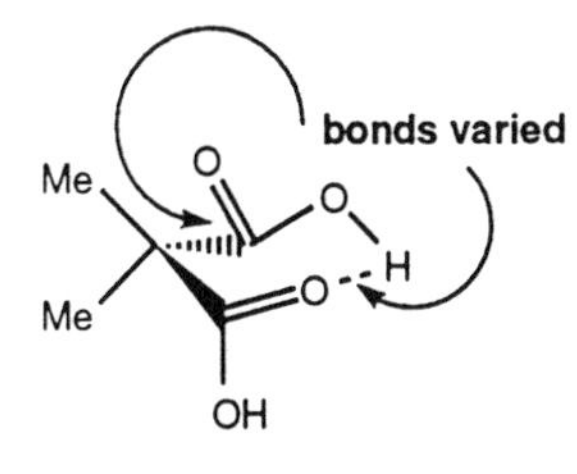

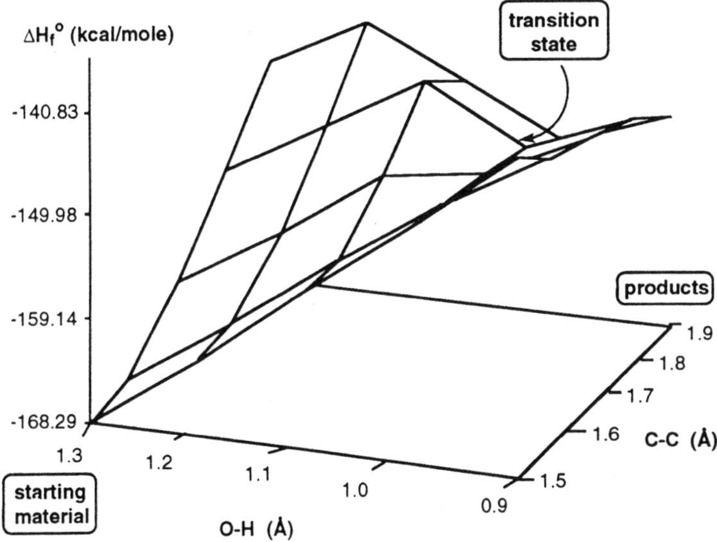

Figure 16.   Part of the reaction surface showing the saddle point.

here made the approach used by Siegel et al.[27] very attractive. In this
approach, two reaction coordinates were systematically varied: (1) the
C–C bond breaking to form CO_2; (2) the O–H bond being formed during
intramolecular proton transfer.

Variation of these two bonds in **29a** by 0.05 Å increments gave a
reaction surface from which a saddle point was located (C–C bond was
varied from 1.5 to 2.0 Å and O–H was varied from 2.0 to 1.0 Å, see Figure
16). Structural refinement gave a geometry which yielded one negative
force constant from the FORCE calculation.

The six-membered transition state located ($H_f^\circ = -148.66$ kcal/mole,
activation energy $E_a = 45.61$ kcal/mole and enthalpy of reaction

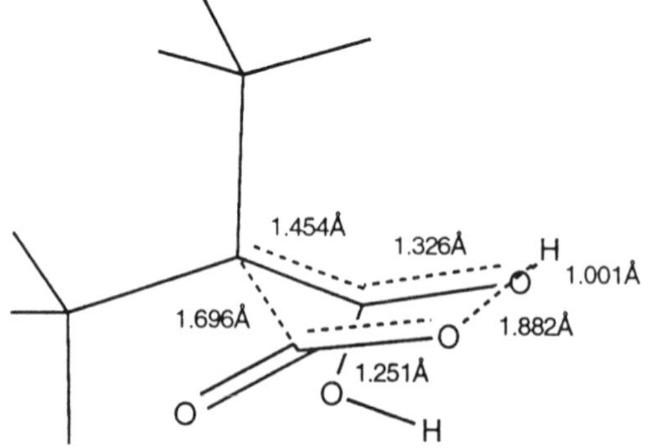

Figure 17. Decarboxylation transition state located for **29a**.

$H_f^\circ{}_{\text{reaction}}$ = 13.98 kcal/mole; see Figure 15 for starting material and product H_f° values) adopted a boat conformation. All attempts to evaluate the alternative chair transition state resulted in rearrangement to the boat conformation. In the transition state the O–H bond being varied adopted a length of 1.001 Å, that is, the proton had almost completely been transferred (see Figure 17). The C–C bond was 0.18 Å longer than in the starting material.

An alternative decarboxylation mechanism considered was the formation of β-lactone **c** from the diacid **a**, which could undergo an acid-catalyzed thermolysis process leading to formation of CO_2 and the ketene acetal **b** (see Figure 18).

The thermolysis of β-lactones was studied by Moyano et al.[28] using MOPAC and implementing the AM1, MNDO and MINDO/3 Hamiltonians. This type reaction was predicted by all three methods to be concerted but highly asynchronous, taking place through a transition state with high zwitterionic character, where all ring atoms lie in the same plane. AM1 calculated E_a and $H_f^\circ{}_{\text{reaction}}$ were closest to the experimental values. The decarboxylation of β-lactones protonated at the carbonyl oxygen atom was also studied as a model for the acid-catalyzed thermolysis process. The system studied was an unsubstituted β-lactone and it was found that the reaction was stepwise and had an overall lower activation energy than the uncatalyzed process (i.e., 37.1 kcal/mole compared to 50.6 kcal/mole).

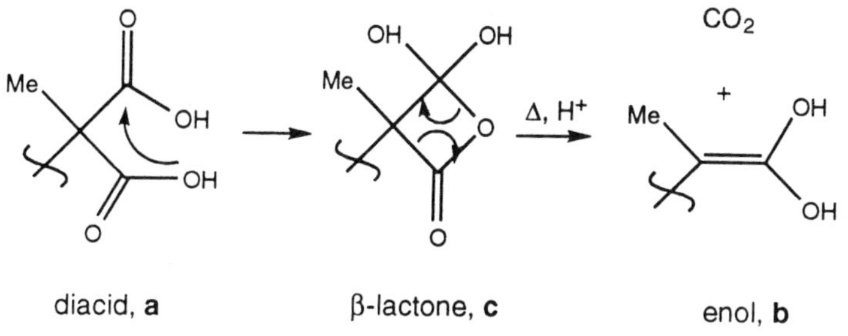

Figure 18. Acid-catalyzed thermolysis of the β-lactone.

Initially, the simple thermolysis transition state for the β-lactone **29c**, resulting from **29a**, was located by applying the transition state bond lengths from the similar molecule studied by Moyano et al. (see Figure 19 and Table 1). The results obtained for **29c** followed the general trend observed by Moyano for the system when $R = H$.

To be able to compare this decarboxylation mechanism with the decarboxylation process via the six-membered transition state, the E_a for formation of the β-lactone **29c** from diacid **29a** had to be determined (see Figure 20). The reaction sequence chosen was suitable for two main reasons: (1) charge was maintained in each of the two reaction going from starting material to products[29]; (2) protonating the carbonyl oxygen also served the purpose of taking into account the fact that experimentally the reaction is acid-catalyzed.

It was not possible to locate the minimum associated with the protonated β-lactone. The initial geometry from PCMODEL, when subjected to minimization in MOPAC fell apart to the protonated diacid

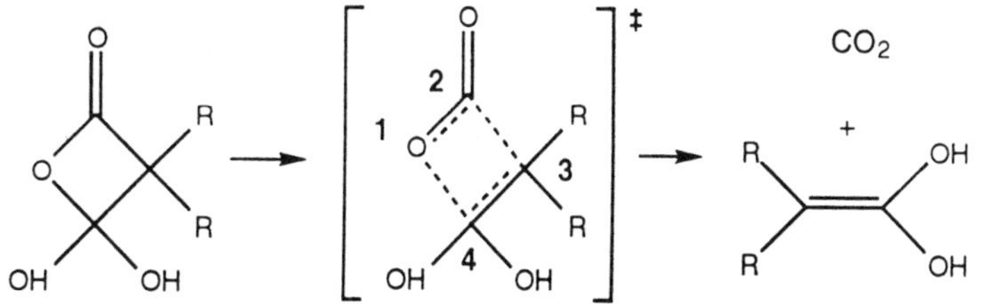

Figure 19. β-Lactone thermolysis studies.

Table 1. Results from the β-Lactone Thermolysis Studies

Transition State Parameters (Å)	*R = H*[b]	*30c, R = Me*[c]
r_{12}	1.289	1.284
r_{23}	1.568	1.589
r_{34}	1.497	1.517
r_{41}	2.155	2.165
H_f° lactone[a]	—[d]	−167.30
H_f° transition state[a]	—[d]	−142.40
H_f° products[a]	—[d]	−180.25
E_a	24.60	24.90
H_f° reaction[a]	−4.80	−12.95

Notes: [a] Units of kcal/mole.
[b] System studied by Moyano.
[c] β-Lactone resulting from cyclization of **30a**.
[d] Results not included in the paper.

starting material. Any attempts to use NLLSQ or SIGMA also resulted in failure; with these alternative techniques the molecule simply fell apart more slowly.

Even in the absence of these results, it was still possible to draw several conclusions about the alternative decarboxylation from the original calculations done on the thermolysis of **29c** (see Figure 19). The reaction profiles for the two alternative decarboxylation processes are shown in Figure 21, where E_a for the formation of the β-lactone **29c** is unknown. Regardless of the value of this unknown E_a, the second step has a transition state with a heat of formation 6 kcal/mole higher than that for

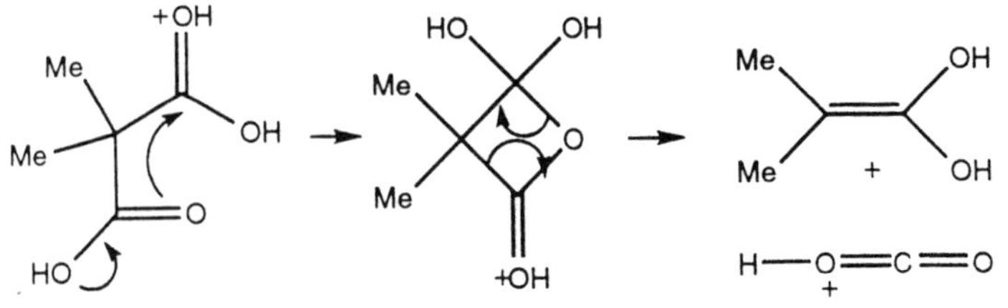

Figure 20. Complete reaction sequence for alternative mechanism.

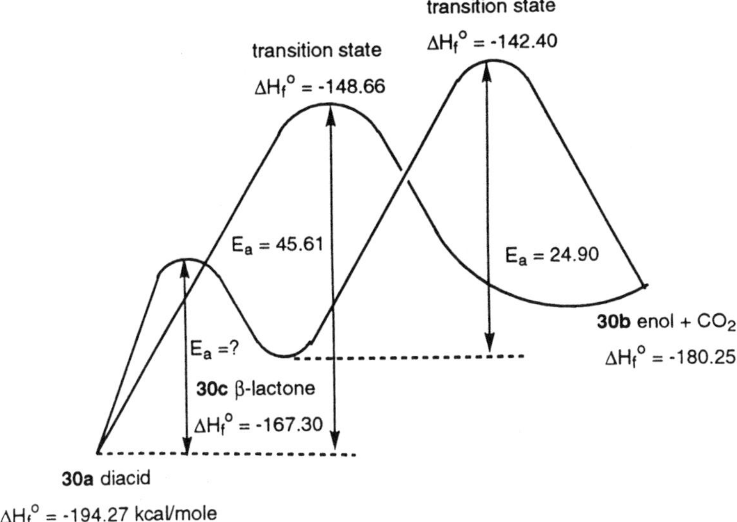

Figure 21. Reaction profiles for the two decarboxylation mechanisms.

the six-membered transition state. Thus, it may be concluded that the decarboxylation process via the six-membered transition state will be energetically more favorable than the alternative process involving β-lactone formation. Having proved the β-lactone mechanism to not be energetically viable for **29a**, it was not investigated with systems **30a** and **23b**.

Model β-Lactam System 30a

There were six possible decarboxylation transition states for the simplified β-lactam **30a**, that is, three separate rotamers about the C–C bond joining the diacid to the β-lactam ring to be considered (C, D and E), each of these having two different conformations depending on which of the two carboxylic acid groups is transferring the proton and being eliminated as CO$_2$ (see Figure 22).

In the Merck rationale of their experimental results, only transition state conformations C1 and D1 were given consideration; C1 being postulated to be lower in energy than D1. The rationale for C1 being the lower energy transition state was because in it, CO$_2$ could be eliminated

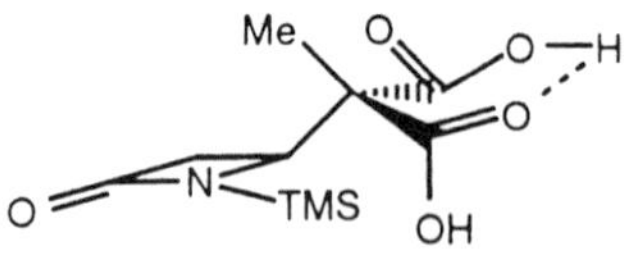

                                    DEBORAH K. JONES and DENNIS LIOTTA

30a conformation C1

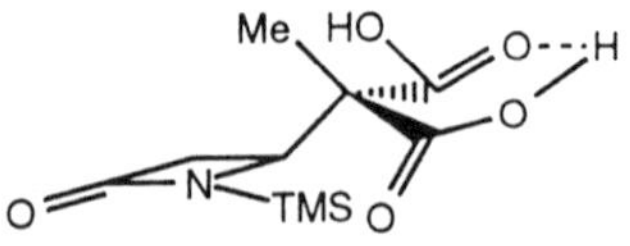

30a conformation C2

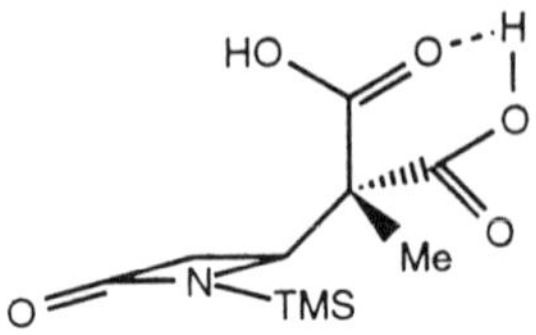

30a conformation D1

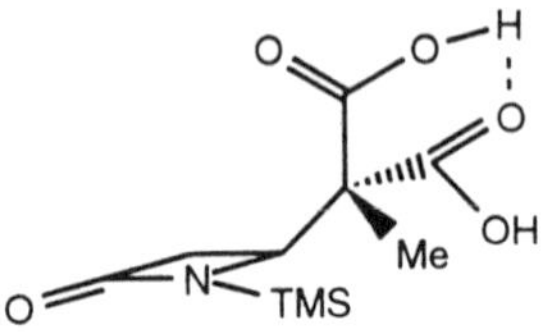

30a conformation D2

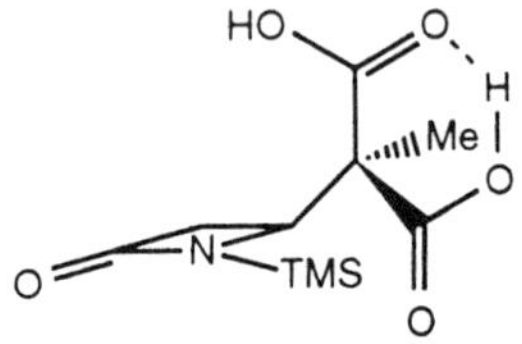

30a conformation E1

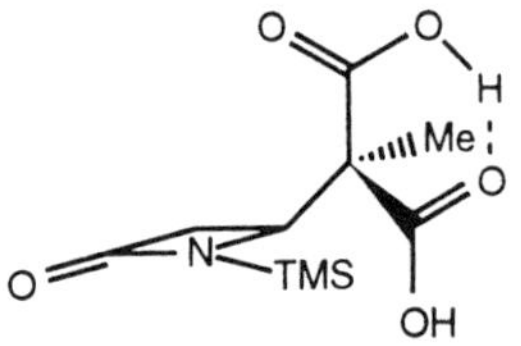

30a conformation E2

Figure 22. Six possible decarboxylation transition states.

unhindered by any other substituents from the back of the molecule. Conversely in D1, CO_2 is eliminated from the same side of the molecule as the TBDMS group, which is apparently a more hindered process.

The barrier to rotation between the three rotamers C, D and E was determined by carrying out a dihedral driver calculation about C–C bond connecting the ring to the diacid (the dihedral which was rotated is shown in Figure 23). **30a** was initially optimized using MOPAC to give a starting conformation in which the two carboxylic acid groups were in the preferred conformation. C, D and E were of similar stability having ΔH_f° of –239.86, –239.96, and –239.85 kcal/mole, respectively. The rotational barriers between the three rotamers were approximately, 3, 4, and 5 kcal/mole, which, at the experimental reaction temperature (50 to 80°C), are barriers that are easily overcome.

The bonds lengths adopted in the six-membered transition state for **29a**, were directly applied to simplified β-lactam diacid **30a**, which enabled the six possible transition states to be located (see Tables 2 and 3). Transition states **30a C1**, **30a D2** and **30a E1** all result in the loss of

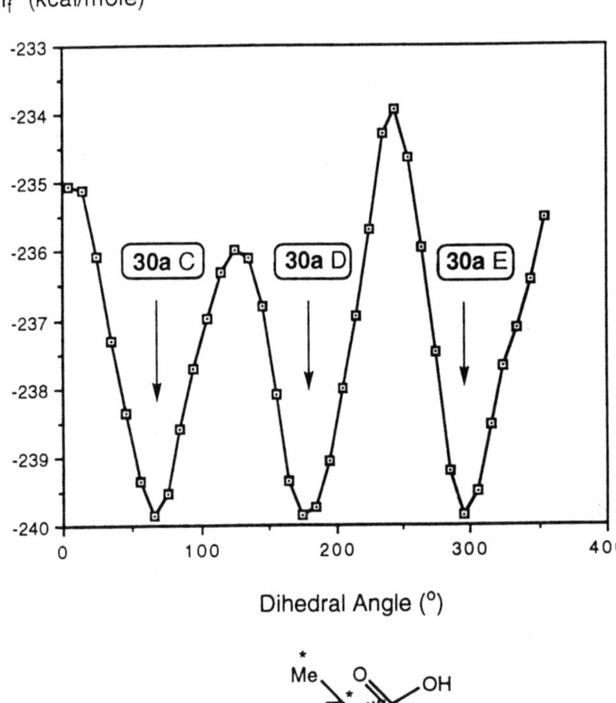

Figure 23. Dihedral driver data for diacid **30a**.

Table 2. MOPAC Results from the Decarboxylation of **30a**

	H_f° (s.m.)^{a,b}	H_f° (t.s.)^{a,c}	H_f° (products)^a	$E_a^{\,a}$	H_f° reaction^a
30a C1	−239.86	−195.91	−230.80	43.95	9.06
30a C2	−239.86	−193.81	−230.26	46.05	9.60
30a D1	−239.96	−196.92	−229.40	43.34	10.56
30a D2	−239.96	−191.83	−230.68	48.13	9.28
30a E1	−239.85	−196.26	−229.62	43.59	10.23
30a E2	−239.85	−193.05	−229.76	46.80	10.09

Notes: ^{a}Units of kcal/mole.

^{b}Starting material diacid minimum for that rotamer.

^{c}Transition state.

Table 3. Amount of Each Carboxylic Acid Group Lost
at Various Reaction Temperatures Determined from
Calculated Partition Functions for **30a**

Temperature (°C)	COOH Experimentally Lost : COOH Retained
50	78.0 : 22.0
60	78.5 : 21.5
70	79.0 : 21.0
80	79.5 : 20.5

the carboxylic acid group that was shown by the ^{13}C labeling studies to be completely lost.

The activation energies (see Table 2) for loss of CO_2 through transition states **30a** C1, **30a** D1 and **30a** E1 were all within 0.6 kcal/mole of each other, which is not in agreement with the Merck postulate of **30a** C1 being the lowest energy transition state. Examination of these three transition states graphically (see Figures 25, 26 and 27) showed that the rationale that the protecting group hinders the loss of CO_2 from that face of the molecule does not fully explain the theoretical results.

The theoretical amount of the labeled carboxylic acid, lost as CO_2 was calculated using the theoretical partition functions (a method which allows for the relative contribution of each transition state to be assessed). From Table 3 it can be seen that the calculations predict that the majority of CO_2 lost is from the same carboxylic acid group that is lost experimentally, but the amount lost falls short by approximately 20% of the experimentally observed 100% loss of only one group. A slight increase in selectivity is predicted theoretically with an increase in temperature.

In an attempt to rationalize the decarboxylation activation energies determined for the six transition states located, the transition state

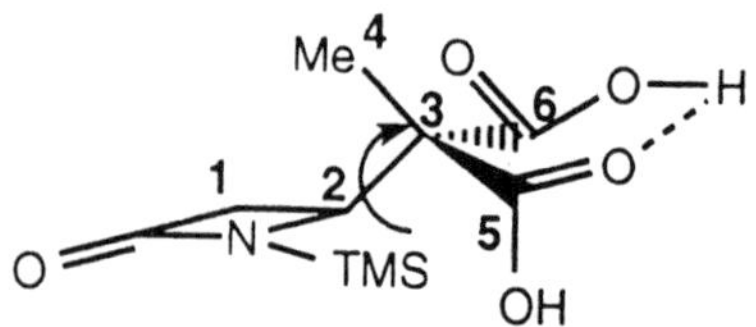

Figure 24. Dihedral angles used to define the position of the dicarboxylic acid.

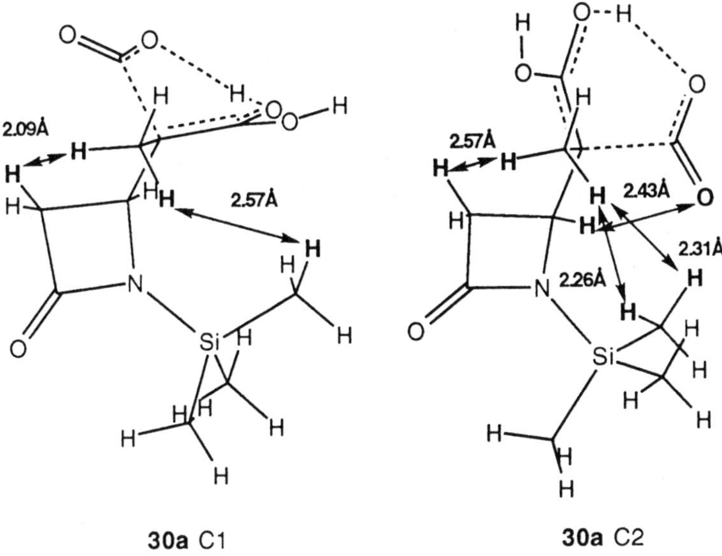

30a C1   **30a** C2

Figure 25. Comparison of the steric interactions in **30a** C1 and **30a** C2.

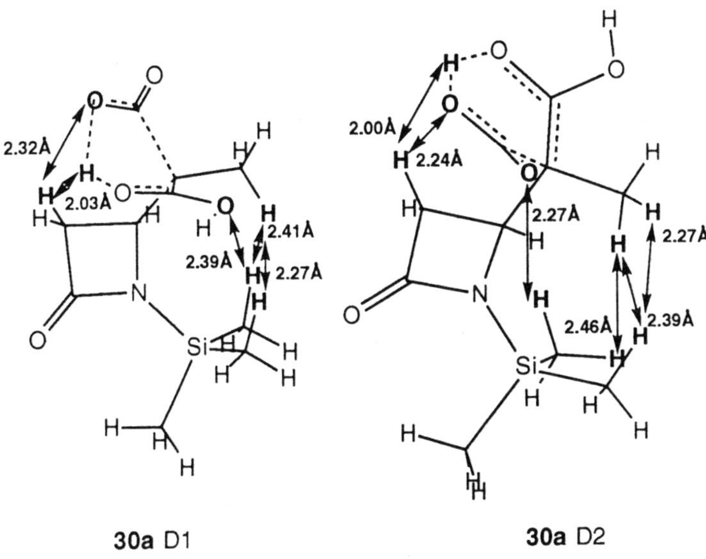

30a D1   **30a** D2

Figure 26. Comparison of the steric interactions in **30a** D1 and **30a** D2.

geometries were examined graphically using SYBYL 6.0. **30a** C1 transition state was more stable than the alternative conformer C2 by 2.10 kcal/mole. Increased interactions between the methyl group and the trimethylsilyl group exist in **30a** C2 compared to C1. Rotation about the C–C bond (see Table 4 and Figure 24) in C2 to optimally position the CO_2 molecule being eliminated results in these interactions (see Figure 25). Overall interactions with the emerging CO_2 molecule and the rest of the molecule are less in C1 than C2.

30a D1 transition state was more stable than the alternative conformer D2 by 4.79 kcal/mole. The molecule of CO_2 being eliminated is subjected to increased steric interactions in transition state **30a** D2 than in **30a** D1 (see Figure 26) which accounts for its decreased stability. **30a** E1 transition state was more stable than the alternative conformer E2 by 3.21 kcal/mole. It is clearly shown in Figure 27 that the CO_2 being eliminated is subjected to increased steric interactions in transition state **30a** E2 than in **30a** E1 (see Figure 27). In addition, interactions between the silyl group and ketene acetal also account for the decreased stability of transition state E2.

In addition to considering steric arguments to rationalize the relative stabilities of the six decarboxylation transition states, the calculated dipole

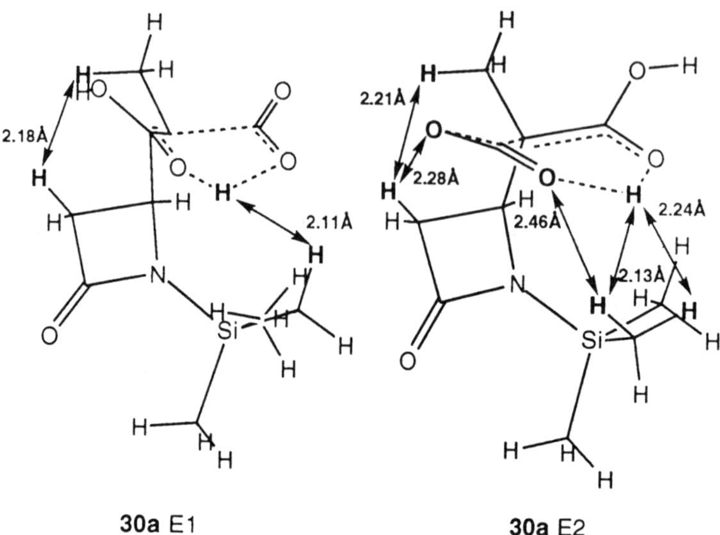

Figure 27. Comparison of the steric interactions in **30a** E1 and **30a** E2.

Table 4. Position of the Diacid Functionality in the Decarboxylation
Transition States for **30a**

Transition State	Dihedral 1-2-3-4[a]	Dihedral 1-2-3-5[a]	Dihedral 1-2-3-6[a]	Dipole Moment[b]
30a C1	56.4	−164.9	−52.4	7.89
30a C2	76.9	−172.3	−62.3	6.37
30a D1	−165.0	−48.6	66.4	6.66
30a D2	−173.6	−61.3	53.2	9.40
30a E1	−60.5	65.8	−177.1	5.38
30a E1	−62.9	56.4	174.9	10.14

Notes: [a]Dihedral angle measured in units of degrees.
     [b]Units of Debye.

moments were examined to see if an electronic argument could be postulated. An overall correlation between calculated dipole moment and activation energy for the six transition states was not observed (see Table 4).

Experimental β-Lactam System 23b

The dihedral driver calculation on the key C–C bond in **23b** (see Figure 28) again showed that the barriers to rotation between C, D and E were

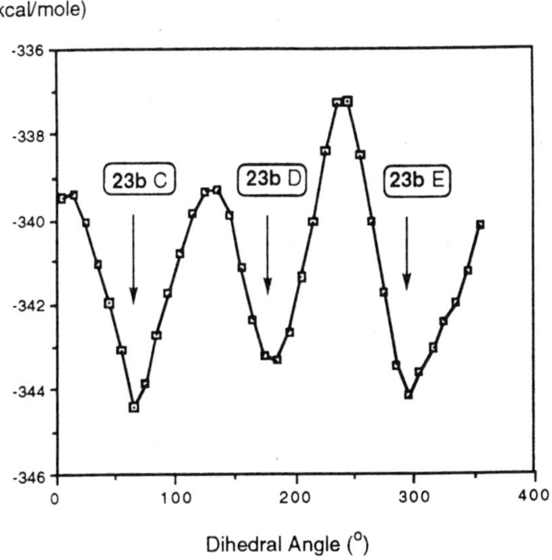

Figure 28. Dihedral driver data for **23b**.

Table 5. MOPAC Results from the Decarboxylation of **23b**

	H_f° $(t.s.)^{a,c}$	H_f° $(s.m.)^{a,b}$	H_f° $(products)^a$	$E_a^{\ a}$	$H_f^\circ{}_{reaction}{}^a$
23b C1	−298.74	−344.38	−331.62	45.64	+9.76
23b C2	−297.19	−344.38	−334.75	47.19	+9.63
23b D1	−299.53	−343.43	−334.89	43.90	+8.54
23b D2	−299.86	−343.43	−333.31	43.57	+10.12
23b E1	−303.95	−344.14	−332.80	40.19	+11.34
23b E2	−296.37	−344.14	−334.23	47.77	+9.91

Notes: [a]Units of kcal/mole.

[b]Starting material diacid minimum for that rotamer.

[c]Transition state.

relatively small, that is, approximately 5, 6, and 7 kcal/mole. There was more variation in the heats of formation than with **30a**: **23b** C, **23b** D and **23b** E ΔH_f° being −344.38, −343.43, and −344.14 kcal/mole, respectively. This increased variance in stability was probably due to the presence of the side chain in this system.

The same approach was taken with **23b** as for **30a**; the bond lengths from **29a** were used to locate the six decarboxylation transition states. The relative stabilities of the transition states located for **23b** were different to those for **30a** (see Table 5). For **30a**, C1, D1 and E1 were all within 0.6 kcal/mole of each other, whereas for **23b**, E1 was 3.38 kcal/mole lower in energy than the next lowest energy transition state.

Modeling the experimental system **23b** gave excellent agreement with the experimental results of 100% loss of only one carboxylic acid group

Table 6. Amount of Each Carboxylic Acid Group Lost at Various Reaction Temperatures Determined from Calculated Partition Functions for **23b**

Reaction Temperature *(°C)*	*COOH Experimentally Lost :* *COOH Retained*
50	99.0 : 1.0
60	98.8 : 1.2
70	98.6 : 1.4
80	98.3 : 1.7

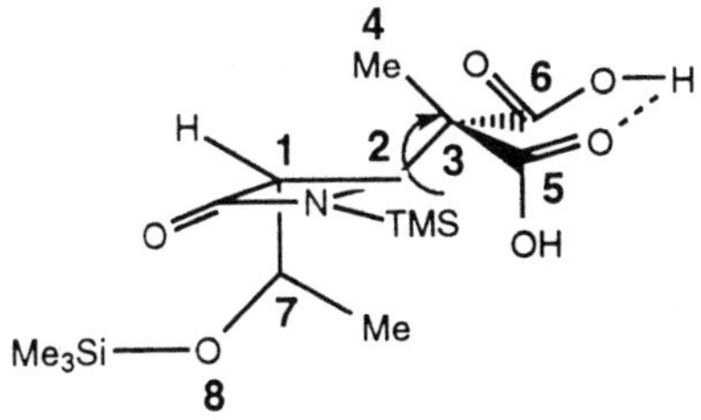

Figure 29. Dihedrals used to define the position of the side chain.

(see Table 6). Although these calculations predicted that the majority of CO_2 lost is from the same carboxylic acid group that is lost experimentally, the lowest energy transition state conformation was **23b** E1, not **23b** C1 as originally hypothesized by the Merck group.

The six decarboxylation transition states associated with **30a** were overlaid in SYBYL 6.0 with the corresponding transition states located for **23b**. This showed that there was very little change in β-lactam ring or diacid conformation going from the **30a** to the **23b** transition states. The values for the three dihedral angles for **30a** listed in Table 4 were almost identical to those in the **23b** transition states.

Two major side chain conformations were observed in the decarboxylation transition states for **23b** (see Table 7). This is illustrated by the two values adopted by dihedral angle **2-1-7-8**. **23b** C1 and **23b** D1 adopted the same side chain conformation, placing the methyl group attached to C_7 (see Figure 29) under the β-lactam ring. The other four remaining

Table 7. Side Chain Conformation for the Decarboxylation Transition States of **23b**

Transition State	Dihedral 2-1-7-8[a]	Dihedral 1-7-8-9[a]	Dipole Moment[b]
23b C1	−160.3	102.9	7.95
23b C2	−70.4	139.0	5.90
23b D1	−162.1	125.7	5.52
23b D2	−58.1	146.5	11.45
23b E1	−74.99	137.40	5.21
23b E1	−63.36	144.87	10.07

Notes: [a]Dihedral angle measured in units of degrees.
[b]Units of Debye.

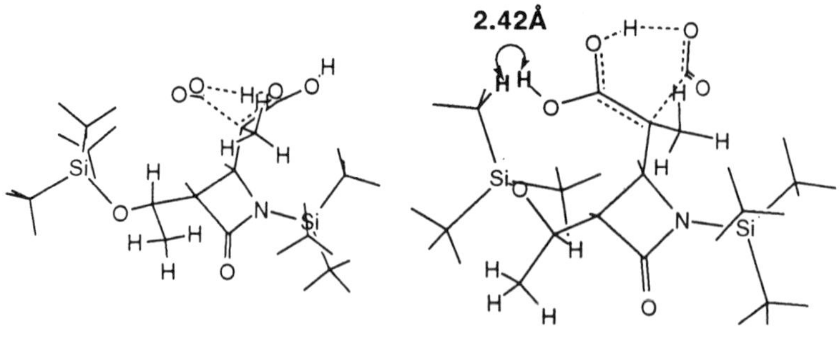

Figure 30. Comparison of the steric interactions in **23b** C1 and **23b** C2.

transition states adopted the alternative conformation, positioning the hydrogen attached to C_7 under the β-lactam ring.

By using SYBYL to graphically analyze the transition state geometries we could demonstrate that the major interactions present in **30a** C1 and C2 (see Figure 25) were also present in **23b** C1 and C2. For **23b** C1, no major interactions occurred between the side chain and the rest of the molecule. The change in side chain conformation from C1 to C2 probably occurred to minimize steric interactions between the methyl groups of the TMSO group and the ketene acetal moiety. Only one interaction was found in **23b** C2 between the side chain and diacid moiety (see Figure 30). This analysis explained why **23b** C1 and C2 have the same relative stabilities as the model β-lactam system transition states **30a** C1 and C2.

23b D1 transition state and the alternative conformer D2 were found to have heats of formation within 0.33 kcal/mole of each other. This is in contrast to the results observed for **30a** D1 and D2, in which D1 was more stable than D2. Interactions present in **30a** D1 were present and increased in **23b** D1. In addition to this, steric interactions existed with the side chain (see Figure 31). Also for **23b**, D2 interactions occurring in **30a** D2 were present and increased. Side chain interactions were minimized by adoption of the alternative conformation in **23b** D2.

For **23b** E1 interactions corresponding to those in **30a** E1 decreased; conversely for **23b** E2, these interactions increased. Both **23b** E1 and E2

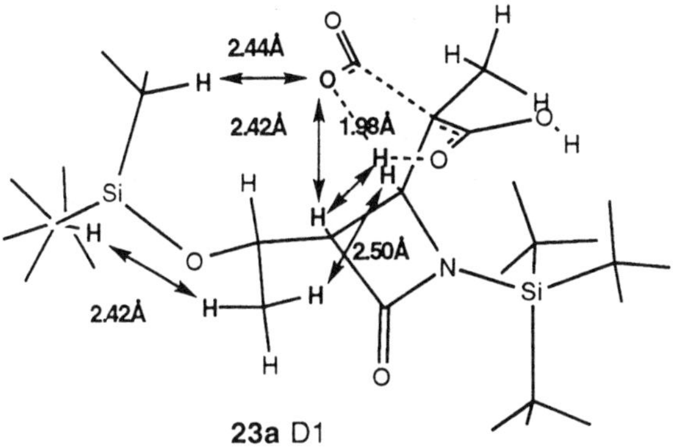

Figure 31. Comparison of the steric interactions in **23b** D1 and **23b** D2.

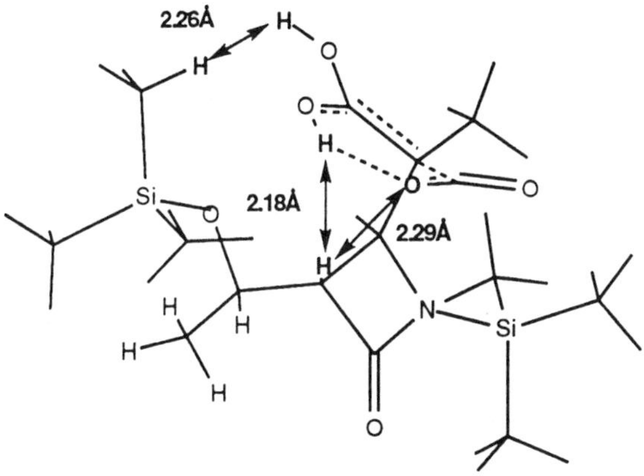

adopted the same side chain conformation. This is because in both of these transition states the group interacting with the side chain is not the ketene acetal or CO_2, but the methyl group. The methyl group due to its branched nature is the most important determinant of the side chain conformation. Adopting the conformation shown in Figure 32 minimizes possible destabilizing interactions. As for **30a**, an overall correlation between calculated dipole moment and activation energy for the six transition states was not observed (see Table 7).

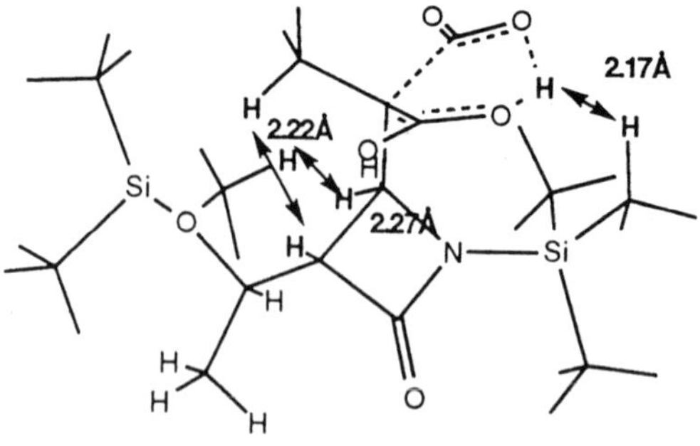

23a E1

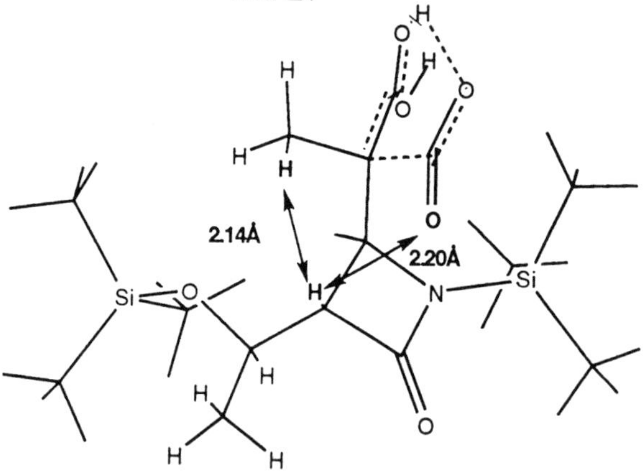

23a E2

Figure 32. Comparison of the steric interactions in **23b** E1 and **23b** E2.

B. Protonation Studies

Model System 29b

The theoretical approach to investigating the ketene acetal protonation
had to fulfill the following criteria: (1) charge had to be maintained going
from starting material to products to prevent the problems which can
occur in such theoretical calculations[29]; (2) the starting materials and
products needed to be computationally-stable species. For example,

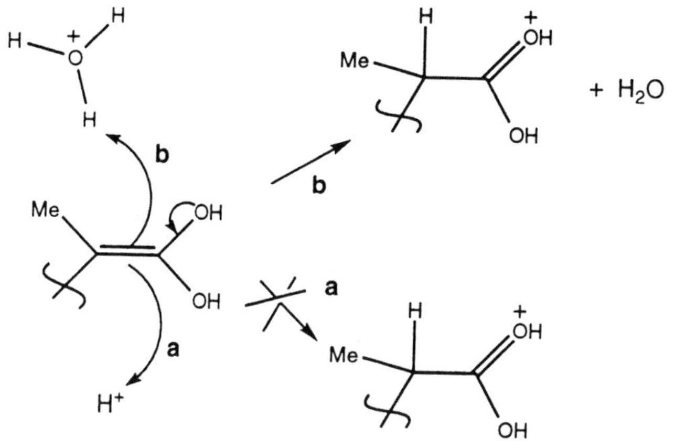

Figure 33. Approaches for studying the protonation.

addition of a proton to the ketene acetal would fulfill the first criteria, but would not fulfill the second, since the "naked" proton would be unstable and readily collapse and add to the ketene acetal double bond (see Figure 33, pathway a).

The approach adopted to study the protonation process was to use a hydronium ion as the proton source, which meet the criteria previously stated (i.e., starting material, H_3O^+, H_2O and protonated carboxylic acid are computationally-stable species). It was reasonable to assume water was present in the formic acid catalyst used in this reaction, thus the hydronium ion species was also a chemically sensible choice (see Figure 33, pathway b).

The protonation reaction surface was generated by taking the product geometry and introducing a water molecule to the system at a distance of 1.00 Å from the proton added to the double bond. Two reaction coordinates, the C–H bond being formed and the O–H bond being broken in the hydronium ion species, were systematically varied in 0.05 Å increments from 1.20 to 2.00 Å and 1.00 to 1.60 Å, respectively. A potential transition state for this process was located as a saddle point on the reaction surface and further structural refinement gave a geometry that was subjected to a FORCE calculation. This calculation yielded two negative force constants, that could not be reduced to a single negative force constant using the IRC procedure.

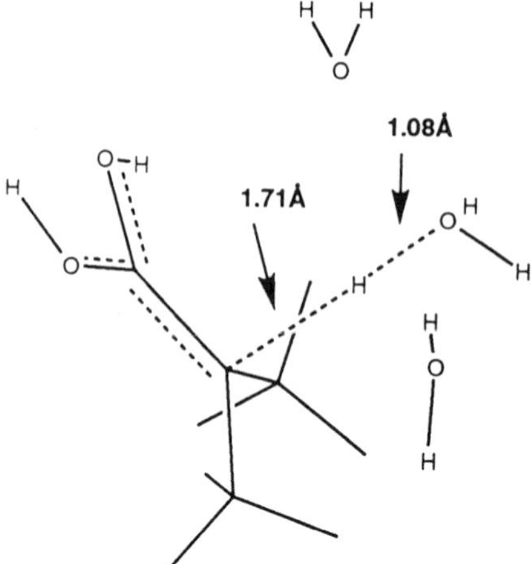

Figure 34. Solvated model protonation transition state geometry.

Examination of this geometry graphically showed that nonbonded interactions were occurring between the hydronium ion and the hydroxy groups of the ketene acetal. These interactions were obviously an artifact of the calculation and were thought to be responsible for the second negative force constant. Experimentally these interactions would never occur, since the process takes place in a solvated environment.

The introduction of solvent molecules was thought to be a reasonable way of preventing the artificial non-bonded interactions and would thus allow the location of the protonation transition state. Two water molecules were added to the system and positioned in such a way as to solvate the hydronium ion species. The solvated system was partially optimized, holding the two reaction coordinates constant. This was followed by full optimization using NLLSQ and SIGMA to refine the transition state geometry. The subsequent FORCE calculation gave one negative force constant, indicating that the model protonation transition state had been located. The key reaction coordinates, C–H and O–H adopted bond lengths of 1.71 Å and 1.08 Å, respectively, in the solvated model protonation transition state geometry (see Figure 34) with the angle of approach of the proton being 94.3°.

To complete the full characterization of the protonation reaction, **29b**, H_2O, H_3O^+ and the protonated methyl carboxylic acid product structures

Table 8. Results from the Protonation Studies
of the **29b**/H_3O^+/$2H_2O$ System

	H_f° *(kcal/mole)*
29b solvated transition state	−124.07
29b[a]	−99.09
H_2O	−59.24
H_3O^+	143.48
$2H_2O/H_3O^+$[b]	−20.35
$2H_2O/H_3O^+$/**29b**[b,c]	−130.78
29b protonated product/$3H_2O$[b,c]	−151.97

Notes: [a]The lowest energy ketene acetal conformer with respect to the hydroxy group conformation (see Figure 34).

[b]These molecules were calculated as one system.

[c]This was the approach taken in all subsequent multiple molecule system calculations.

were all optimized separately (see Table 8). Interestingly, the sum of the heats of formation of the four starting material molecules, −74.07 kcal/mole, was less than that for the transition state structure of −124.07 kcal/mole. Calculation of the water molecules and hydronium ion as a complete system and adding this to the heat of formation of **29b**, still gave a total heat of formation (−119.44 kcal/mole) less than that for the transition state. The final solution to this problem was to calculate all the molecules involved in this process as one complete system, for both starting materials and products, thus allowing direct comparison with the transition state system. The resulting E_a for this reaction was 6.71 kcal/mole.

In addition to fulfilling the original criteria set for the computational proton source, use of a hydronium ion had the advantage of not introducing many additional degrees of freedom to the system being studied. Use of a formic acid molecule as the proton source would give a more faithful representation of the experimental conditions, but this had the disadvantage of greatly increasing the degrees of conformational freedom associated with the system.

In an attempt to validate the use of a hydronium ion, rather than formic acid as the proton source, the transition state was located for the protonation of **29b** using protonated formic acid, solvated by two formic acid

Table 9. Comparison of Solvated H_3O^+ and $HCOOH_2^+$ Proton Sources

	β29b/H_3O^+/2H_2O	29b/$HCOOH_2^+$/2HCOOH
ΔH_f° (starting materials)[a]	−130.78	−235.91
ΔH_f° (transition state)[a]	−124.07	−226.42
ΔH_f° (products)[a]	−151.97	−261.89
E_a[a]	6.71	9.49
ΔH_f° reaction[a]	−21.19	−25.98

Note: [a]Units of kcal/mole.

molecules (see Table 9). The similarity between the geometries of the two transition states and the relative energetics of the two systems confirmed that our assumption was valid and that a hydronium ion would act as a good computational surrogate of the experimental proton source.

An alternative, but highly improbable protonation mechanism involved the [1,3]-H shift from one of the ketene acetal hydroxy groups to the carbon center that becomes protonated (see Figure 35). Thermal [1,3]-H rearrangements are symmetry-allowed only as antarafacial shifts, but geometrical constraints prevent them from occurring; suprafacial [1,3] shifts are symmetry-disallowed. For these reasons very few thermal uncatalyzed [1,3] shifts are known (those that are known have very high activation energies).[30]

To be able to discount this mechanistic pathway from the protonation studies, it was necessary to locate the transition state to determine the activation energy for this process. The reaction surface was produced by taking **29b** and varying the O–H bond being broken from 1.00 to 1.60 Å and the C–H bond being made from 1.80 to 1.00 Å in increments of 0.05 Å. Geometry refinement of the saddle point gave a structure which had one negative force constant associated with it when a FORCE calculation

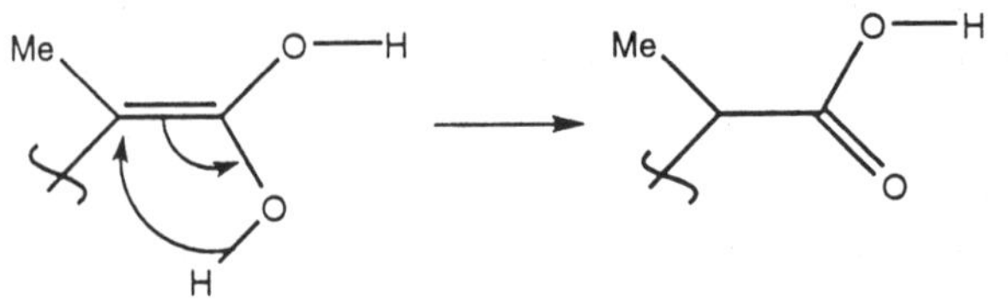

Figure 35. Alternative [1,3]-H shift protonation mechanism.

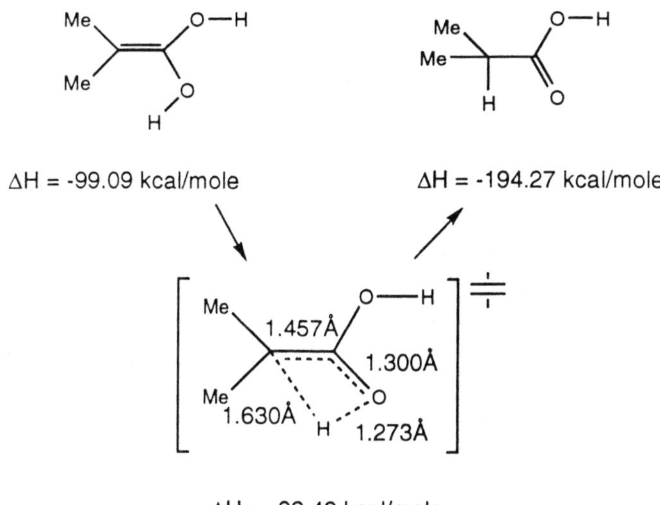

Figure 36. [1,3]-H shift pathway for **29b**.

was computed. The activation energy for this [1,3]-H shift was found to be very high at 66.60 kcal/mole (see Figure 36, Table 10).

These results showed the [1,3]-H shift pathway to be a high energy process, due to the rotation which must occur about the central double bond to orient the proton which is shifting into a suitable conformation. For direct comparison with the original protonation process considered, a hydronium and two water molecules had to be added to the [1,3]-H

Table 10. Comparison of Intra- and Intermolecular Protonation Processes

	Intramolecular [1,3]-H Shift		Intermolecular
	29b	**$29b/H_3O^+/2H_2O$**	**$29b/H_3O^+/2H_2O$**
H_f° (starting materials)[a]	−99.09	−130.78	−130.78
H_f° (transition state)[a]	−32.49	−64.19	−124.07
H_f° (products)[a]	−194.27	−134.78	−151.97
E_a[a]	66.60	66.59	6.71
$H_{f\ reaction}^\circ$[a]	−95.18	4.00	−21.19

Note: [a]Units of kcal/mole.

shift transition state geometry. These three molecules were placed approximately 3 Å from the ketene acetal, and were allowed to optimize whilst the ketene acetal was held fixed in its transition state arrangement. Using NLLSQ and SIGMA, the geometry was refined to locate the transition state geometry (see Table 10).

The E_a for the [1,3]-H shift process, including the hydronium ion and two water molecules, was very high (approximately 60 kcal/mole higher than the E_a for the intermolecular process in which the proton source was a hydronium ion solvated by two water molecules). This study proved that the [1,3]-H shift was not an energetically viable alternative mechanistic pathway to the original mechanism investigated. For this reason the [1,3]-H shift mechanism was not investigated for the other systems, **30b** and **26**.

Model β-Lactam System 30b

Using the results from the model study of **29b** (which used a solvated hydronium as a proton source), protonation of **30b** (the simplified β-lactam) was studied. The four possible transition states were located

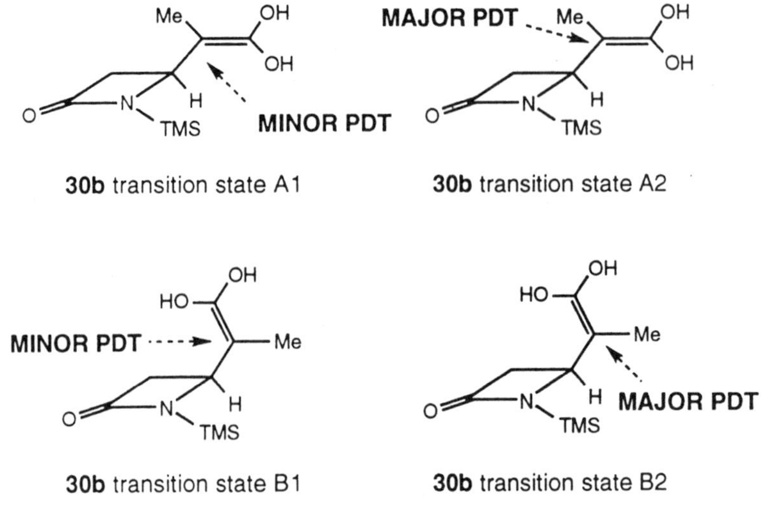

Figure 37. The four possible protonation transition states.

Table 11. Results from the Protonation Study of **30b**

	H_f° (starting materials)[a]	H_f° (transition state)[a]	$E_a^{\,a}$
$H_2O + H_3O^+$			
30b A1[b]	−124.94	−99.83	25.11
30b A2[c]	−124.94	−102.27	22.67
30b B1[b]	−124.66	−102.97	21.69
30b B2[c]	−124.66	−97.53	27.13
$2 H_2O + H_3O^+$			
30b A1[b]	−198.8	−171.26	27.54
30b A2[c]	−198.8	−169.48	29.32
30b B1[b]	−198.43	−173.77	24.66
30b B2[c]	−198.43	−167.3	31.13
$4H_2O + H_3O^+$			
30b A1[b]	−334.61	−305.26	29.35
30b A2[c]	−334.61	−305.57	29.04
30b B1[b]	−334.08	−308.29	25.97
30b B2[c]	−334.08	−302.62	31.46

Notes: [a]Units of kcal/mole.

[b]Leads to the major product observed experimentally.

[c]Leads to the minor product observed experimentally.

(see Figure 37) by applying the pertinent bond lengths, bond angles and dihedral angles determined from the model study. This system was studied at three levels of hydronium ion solvation, that is, using one, two and four water molecules. This was done to confirm that using the arbitrary number of two water molecules originally was a valid approach.

Table 12. Product Distribution Determined from Calculated Partition Functions

	Calculated Product Distributions at Various Reaction Temperatures (°C)			
	50	60	70	80
$H_2O + H_3O^+$	99.4 : 0.6	99.3 : 0.7	99.1 : 0.9	99.0 : 1.0
$2H_2O + H_3O^+$	93.6 : 6.4	93.5 : 6.5	93.4 : 6.6	93.4 : 6.6
$4H_2O + H_3O^+$	90.2 : 9.8	88.4 : 11.6	86.6 : 13.4	84.4 : 15.6

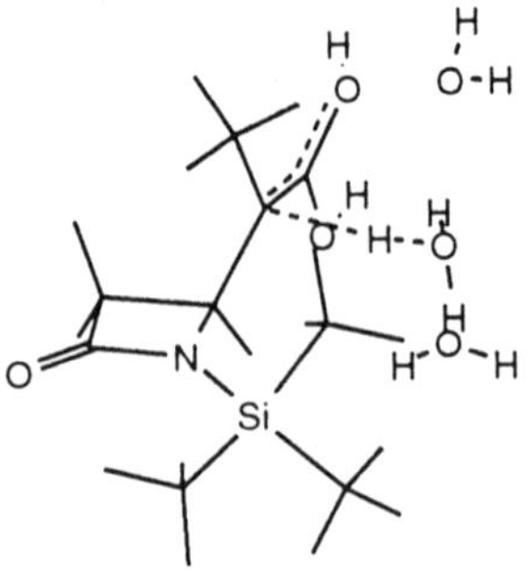

30a A1, E_a = 27.54 kcal/mole 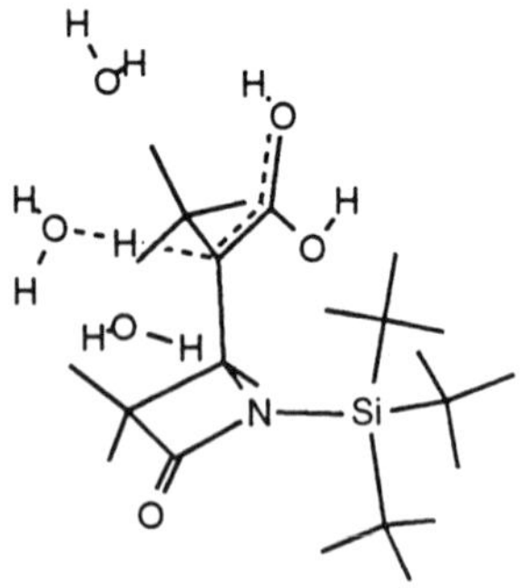　　　　**30a** A2, E_a = 29.32 kcal/mole

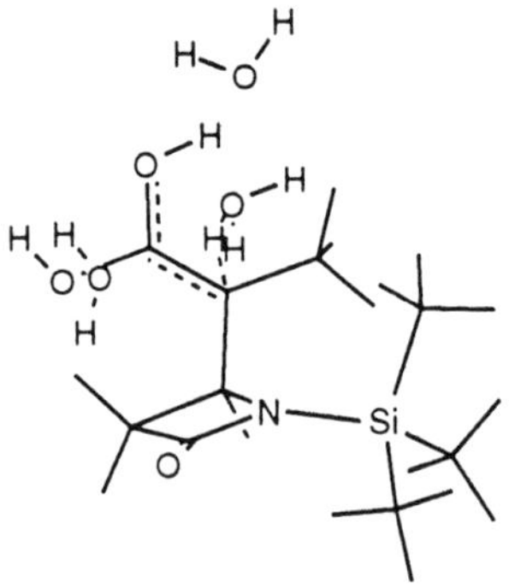

30a B1, E_a = 24.66 kcal/mole 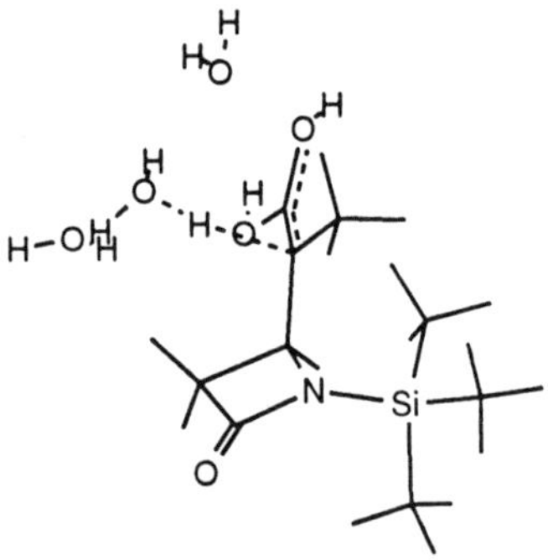　　　　**30a** B2, E_a = 31.13 kcal/mole

Figure 38. **30a** protonation transition states located with two molecules of water present solvating the hydronium ion proton source.

At all three levels of solvation the four protonation transition states had the same relative stabilities and in all the systems studied the lowest activation energy was for the protonation process via transition state **30b** B1 (see Table 11).

Using the results shown in Table 11 and the partition functions calculated during the FORCE calculation, the product distribution was determined for the systems studied at a variety of reaction temperatures (see Table 12). The calculations correctly predicted formation of the isomer that has been shown to form experimentally as the major product. Also the calculated product distributions were in reasonable agreement with

Table 13. Comparison of Activation Energies and Dipole Moments

	E_a (kcal/mole)	Dipole Moment (Debye)
$H_2O + H_3O^+$		
30b A[a]	—	9.54
30b B[a]	—	8.2
30b A1[b]	25.11	12.47
30b A2[c]	22.67	13.93
30b B1[b]	21.69	6.78
30b B2[c]	27.13	15.86
$2\,H_2O + H_3O^+$		
30b A[a]	—	10.53
30b B[a]	—	9.35
30b A1[b]	27.54	12.15
30b A2[c]	29.32	14.66
30b B1[b]	24.66	8.89
30b B2[c]	31.13	14.44
$4H_2O + H_3O^+$		
30b A[a]	—	6.04
30b B[a]	—	5.37
30b A1[b]	29.35	11.97
30b A2[c]	29.04	11.98
30b B1[b]	25.79	9.13
30b B2[c]	31.46	11.90

Notes: [a]Starting materials.

[b]Leads to the major product observed experimentally.

[c]Leads to the minor product observed experimentally.

the experimentally observed results (i.e., $\beta{:}\alpha_{exp}$ observed ranging from 96:4 to 94:6 for reaction temperatures from 50 to 80°C).

Although the formation of the experimentally observed major product had been predicted by the theoretical studies, the lowest energy transition state was not in agreement with the transition state which had been proposed by the Merck group to rationalize the experimental results observed. As stated earlier, **30b** A1 was postulated as being the most favorable transition state because: (1) conformation A was expected to be preferred due to the unfavorable A-1,3 strain between the ketene acetal hydroxy and the C_3 methine proton in conformation B and (2) protonation was expected to occur from the face opposite to the bulky *N*-silyl protecting group.

In an attempt to understand the computational results obtained, the protonation transition states, A1, A2, B1 and B2, were examined graphically using SYBYL (see Figure 38).

From the graphical examination of the transition states shown in Figure 38, the origin of the predicted selectivity was not apparent. No bad steric interactions were located that would significantly favor one transition state over another. Interestingly, it was found that in the lowest energy transition state **30a** B1, the ketene acetal functionality was not positioned directly above the β-lactam ring. Instead it was rotated to a conformation placing it perpendicular to the β-lactam ring and thus eliminating the A-1,3 strain predicted to be associated with conformation B.

In the absence of a rationale invoking steric arguments, the calculated dipole moments of the transition states located were examined to see if they could provide an alternative rationale (see Table 13). For the three solvated systems investigated, the lowest energy transition state B1 in all cases also had the lowest dipole moment, suggesting a correlation between dipole and transition state stability. The consequence of these results is still not fully understood.

The barrier to rotation about the C–C bond joining the β-lactam ring to the ketene acetal was determined by a dihedral driver calculation around this bond. It was of great importance to know the energy required to rotate about this bond and convert conformation A to conformation B

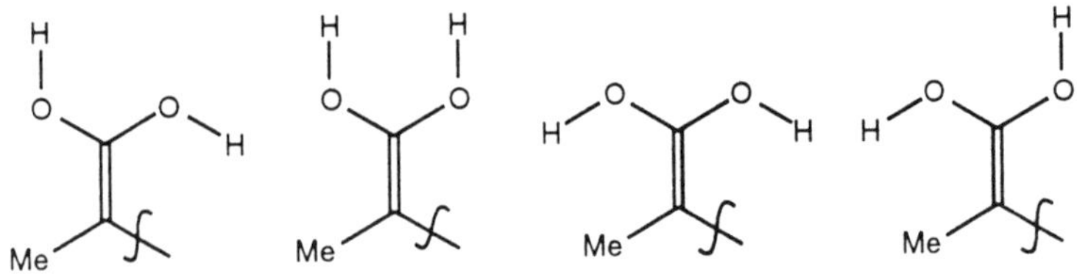

Figure 39. Possible ketene acetal conformations.

(see Figure 10 and 12) which was proposed to be less stable due to A-1,3 strain. Before carrying out the dihedral calculation, the most stable ketene acetal conformation was determined for both **30b** A and **30b** B. The possible conformations are shown in Figure 39.

The most stable ketene acetal conformation for **30a** A and **30a** B was IV (see Table 14) and this was the conformation used for the dihedral driver calculation. There was a very low barrier to rotation (approximately 2.6 kcal/mole between **30b** A and **30b** B). This proved that even if the decarboxylation is specific, occurring from only one diacid conformation to give selectively one ketene acetal conformation, this is irrelevant, since rotation between A and B must be facile.

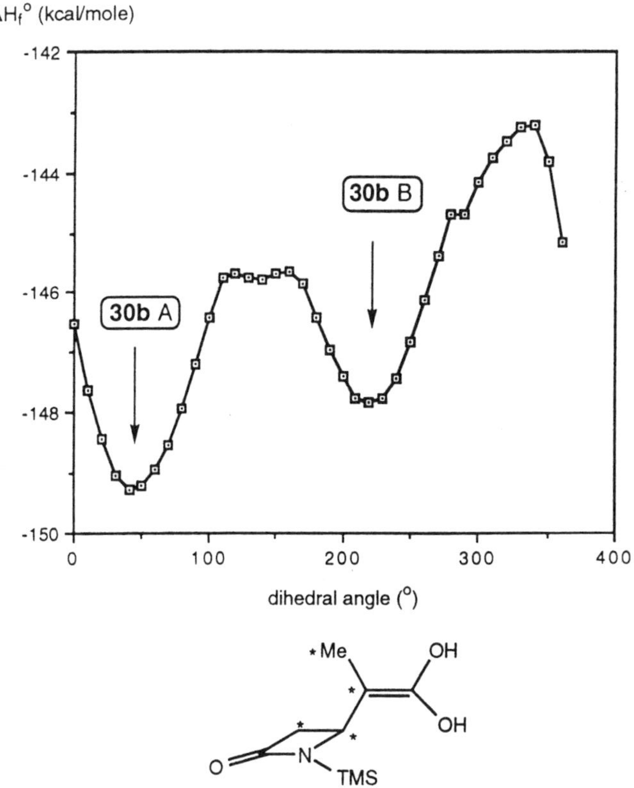

Figure 40. Dihedral driver data for **30b**.

Table 14. Results of MOPAC Calculations for Various Ketene Acetal Conformations

	ΔH_f° For Ketene Acetal Conformations[a] (see Figure 39)			
	I	II	III	IV
30a A	−147.52	−149.28[b]	−144.17	−149.28
30a B	−147.57	−147.86[b]	−141.86	−147.86

Notes: [a]Units of kcal/mole.

[b]Conformation II rotated to IV during minimization.

Experimental β-Lactam System 26

The barrier to rotation about the C–C bond joining the β-lactam ring to the ketene acetal **26**, was determined by a dihedral driver calculation around this bond. As for **30b**, before carrying out the dihedral calculation the most stable ketene acetal conformation was determined for both **26** A and **26** B (see Figure 39).

The most stable ketene acetal conformation for **26** A and **26** B, was IV (see Table 15) and this was the conformation used for the dihedral driver calculation. There was a very low barrier to rotation, approximately 5 kcal/mole, between **26** A and **26** B (see Figure 41). Again, as for the model β-lactam system **30b**, this proved that it is the kinetic protonation of the ketene acetal that determines the final product distribution regardless of the specific decarboxylation step occurring.

Table 15. Results of MOPAC Calculations for Various Ketene Acetal Conformations

	ΔH_f° For Ketene Acetal Conformations[a] (see Figure 39)			
	I	III	II	IV
26 A	−252.00	−253.31[b]	−249.08	−253.31
26 B	−251.56	−252.87[b]	−248.78	−252.87

Notes: [a]Units of kcal/mole.

[b]Conformation II rotated to IV during minimization.

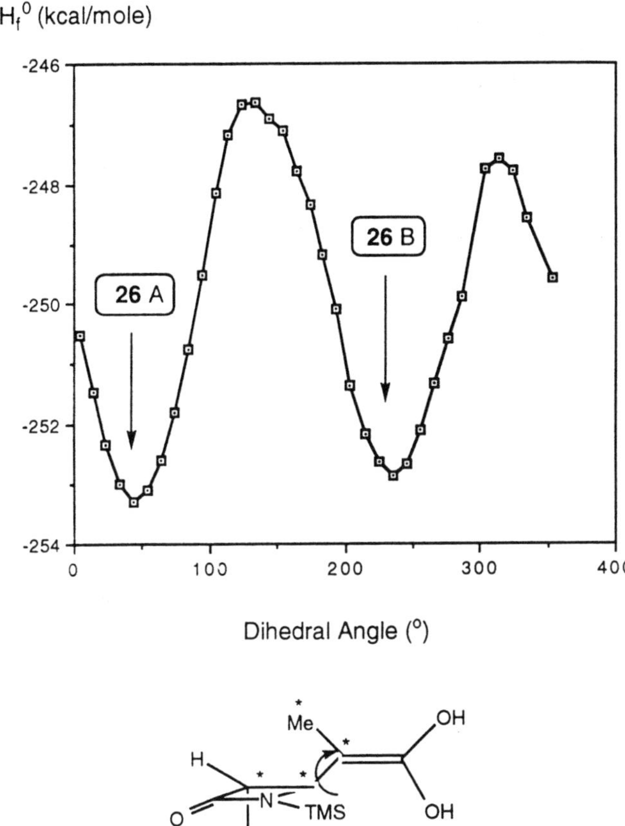

Figure 41. Dihedral data for **26**.

IV. CONCLUSIONS

Importantly, all the calculations predicted the formation of the experimentally observed major product. In a general sense, the theoretical results were in agreement with the mechanism proposed by Choi et al. which postulated the occurrence of both diastereospecific decarboxylation and selective protonation processes.

The difference between the original mechanism and the mechanism that was subsequently suggested as a consequence of the findings of the

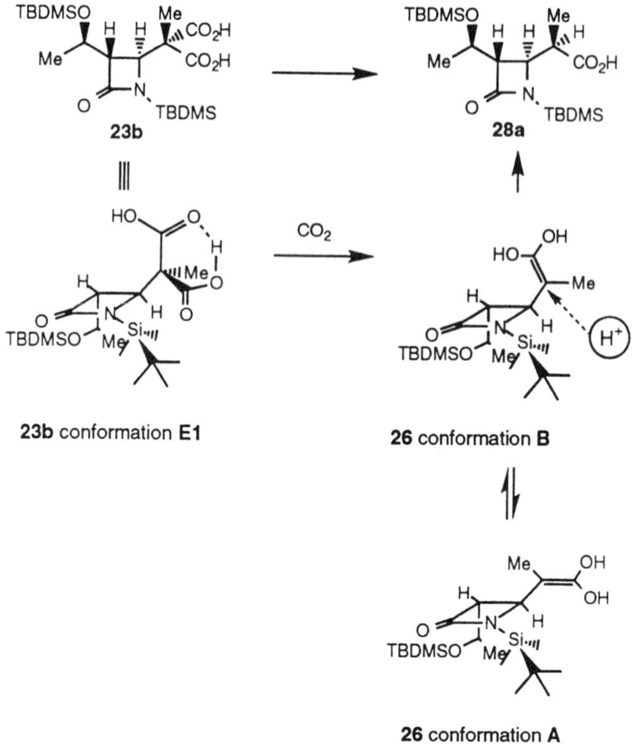

Figure 42. Mechanism proposed on the basis of the calculations.

theoretical investigations lies in the conformation adopted in the lowest energy transition states for both decarboxylation and protonation reactions (see Figure 42).

For the decarboxylation process, conformation **E1**, in which CO_2 is lost from the same face of the molecule that the *N*-silyl protecting group occupies was found to be more than 3 kcal/mole lower in energy than the next lowest transition state. This is in contrast with the original postulate of a transition state adopting conformation **C1** providing the lowest energy decarboxylation pathway (see Figure 10 for the original Merck mechanism). This suggests that the *N*-silyl protecting group does not significantly effect loss of CO_2 in transition state **E1**.

For the protonation process, a transition state in which the ketene acetal adopting conformation **B**, with subsequent protonation from the same face of the molecule that the *N*-silyl protecting group occupies was found to give the lowest energy protonation pathway. This also is in disagreement with the protonation conformation that was originally proposed as being possibly energetically favorable, that is, conformation **A** with protonation from the face opposite to the *N*-silyl protecting group. The unfavorable A-1,3 strain in **B** was relieved by rotating the ketene acetal so it was not directly positioned above the β-lactam ring and the effect of the *N*-silyl protecting group was not sufficient to prevent protonation of that face of the double bond.

Of great interest was the barrier to rotation between conformation **A** and **B** in the ketene acetal systems studied. In all cases this was shown to be very small. From this, the final conclusion can be drawn that even if the decarboxylation is specific, the key step in this reaction is the selective protonation of the ketene acetal. This is what determines the final product distribution. The fact that the decarboxylation step is highly selective just happens to be an interesting coincidence.

REFERENCES

1. (a) Shih, D.H.; Baker, F.; Cama, L.; Christensen, B.G. *Heterocycles*, **1984**, *21*, 29. (b) Shih, D.H.; Fayter, J.A.; Cama, L.D.; Christensen, B.G. *Tetrahedron Lett.* **1985**, *26*, 583.
2. Shih, D.H.; Cama, L.; Christensen, B.G. *Tetrahedron Lett.* **1985**, *26*, 587.
3. Shibata, T.; Iino, K.; Tanaka, T.; Hashimoto, T.; Kameyama, Y.; Sugimura, Y. *Tetrahedron Lett.* **1985**, *26*, 4739.
4. Iimori, T.; Shibasaki, M. *Tetrahedron Lett.* **1986**, *27*, 2149.
5. Kitamura, M.; Nagai, K.; Hsiao, Y.; Noyori, R. *Tetrahedron Lett.* **1990**, *31*, 549.
6. Uyeo, S.; Itani, H. *Tetrahedron Lett.* **1991**, *32*, 2143.
7. Martel, A.; Daris, J.P.; Bachand, C.; Corbeil, J.; Menard, M. *Can. J. Chem.* **1988**, *66*, 1537.
8. Fuentes, L.M.; Shinkai, I.; Slazmann, T.N. *J. Am. Chem. Soc.* **1986**, *108*, 4675.
9. Fuentes, L.M.; Shinkai, I.; King, A.; Purick, R.; Reamer, R.A.; Schmitt, S.M.; Cama, L.; Christensen, B.G. *J. Org. Chem.* **1987**, *52*, 2563.
10. Bender, D.R.; DeMarco, A.M.; Melillo, D.G.; Riseman, S.M.; Shinkai, I. *J. Org. Chem.* **1992**, *57*, 2411.
11. Nagao, Y.; Kumagai, T.; Tamai, S.; Abe, T.; Kuramoto, Y.; Taga, T.; Aoyagi, S.; Nagase, Y.; Ochiai, M.; Inoue, Y.; Fujita, E. *J. Am. Chem. Soc.* **1986**, *108*, 4673.
12. Prasad, J.S.; Liebeskind, L.S. *Tetrahedron Lett.* **1987**, *28*, 1857.
13. Kawabata, T.; Kimura, Y.; Ito, Y.; Terashima, S.; Sasaki, A.; Sunagawa, M. *Tetrahedron* **1988**, *44*, 2149.

14. Ito, Y.; Sasaki, A.; Tamoto, K.; Sunagawa, M.; Terashima, S. *Tetrahedron* **1991**, *47*, 2801.
15. Bender, D.R.; DeMarco, A.M.; Melillo, D.G.; Riseman, S.M.; Shinkai, I. *J. Org. Chem.* **1992**, *57*, 2411.
16. Choi, W.B.; Churchill, H.R.O.; Lynch, J.E.; Thompson, A.S.; Humphrey, G.R.; Volante, R.P.; Reider, P.J.; Shinkai, I. In press.
17. We are grateful to Kevin Gilbert and Mark Midland for making these programs available to us.
18. Gilbert, K.; Gajewski, R. In *Advances in Molecular Modeling*, Vol. 2, D. Liotta (ed.), JAI Press, Greenwich, CT, 1990.
19. (a) Stewart, J.J.P. *J. Comp. Aided Mol. Des.* **1990**, *4*, 1. QCPE Program No. 455; (b) Jones, D.K.; Liotta, D.C.; Choi, W-B.; Volante, R.P.; Reider, P.J.; Shinkai, I.; Churchill, H.R.O.; Lynch, J.E. *J. Org. Chem.* **1994**, *59*, 3749.
20. (a) Dewar, M.J.S.; Zoebisch, E.G.; Healy, E.F.; Stewart, J.J.P. *J. Am. Chem. Soc.* **1985**, *107*, 3902. (b) Dewar, M.J.S.; Jie, C. *Organometallics* **1987**, *6*, 1486.
21. Truhlar, D.G.; Steckler, R.; Gordon, M.S. *Chem. Rev.* **1987**, *87*, 217.
22. (a) Goddard, J.D.; Yamaguchi, Y.; Schaefer III, H.F. *J. Chem. Phys.* **1992**, *96*, 1158. (b) Francisco, J.S. *J. Chem. Phys.* **1992**, *96*, 1167.
23. Nguyen, M.T.; Ruelle, P. *Chem. Phys. Lett.* **1987**, *138*, 486.
24. Murto, J.; Raaska, T.; Kunttu, H.; Rasanen, M. *J. Mol. Struct. (Theochem)* **1989**, *200*, 93.
25. Bock, C.W.; Redington, R.L. *J. Chem. Phys.* **1986**, *85*, 5391.
26. Bock, C.W.; Redington, R.L. *J. Phys. Chem.* **1988**, *92*, 1178.
27. Siegel, M.M.; Colthup, N.B. *Appl. Spectrosc.* **1987**, *41*, 1227.
28. Moyano, A.; Pericas, M.A.; Valenti, E. *J. Org. Chem.* **1989**, *54*, 573.
29. Maintaining charge is discussed in great detail in Chapter 2.
30. Woodward, R.B.; Hoffman, R. *The Conservation of Orbital Symmetry*, Verlag Chemie Academic Press, Weinheim, Germany, 1970.

MOLECULAR MODELING OF CARBOHYDRATES

D. Ross Boswell, Edward E. Coxon, and

James M. Coxon

Advances in Molecular Modeling
Volume 3, pages 145–193.
Copyright © 1995 by JAI Press Inc.
All rights of reproduction in any form reserved.
ISBN: 1-55938-326-7

I.   THE FUNCTION OF COMPLEX CARBOHYDRATES

A.   Introduction

Carbohydrate chemistry is in some ways both the oldest and the newest branch of organic chemistry. Although Emil Fischer's research on carbohydrates in the late nineteenth century was one of the seminal events in natural product chemistry, it is only in the last two decades that the ubiquitous carbohydrate molecules which are attached to proteins and lipids and coat cellular surfaces, have come to be studied as biochemically important in their own right. We briefly review some of the processes in which carbohydrates play a central role. The short history of carbohydrate modeling and its effect on the current conceptualization of biochemical processes involving carbohydrates is discussed. Molecular modeling studies in conjunction with NMR (nOe) methods provide a powerful tool to understand these important biochemical processes.

The precise biological functions and mechanisms of action of carbohydrates are unknown; however, in recent years considerable information on their biosynthesis has been accumulated. The resultant scheme, consisting of a remarkably complicated sequence of specific anabolism followed by an equally specific sequence of programmed catabolism and resynthesis, demands too much biochemical effort to be devoid of function.[1] The diverse functions of carbohydrates can be broadly classified as cellular recognition processes, placing them at the heart of one of the mysteries central to biological systems. For example, cellular recognition is central to the processes of fertilization, embryogenesis, cellular organization into specific tissues,[2] neural development, hormonal activities, cell proliferation, inflammation, blood group incompatibility, a variety of diseases through antibody–antigen recognition, and viral and bacterial attachment to host cells, possibly including infection by HIV.[3] In addition, some carbohydrates act as tumor agents.

Two major developments in the late 1960s drew attention to the concept of carbohydrate mediated cellular recognition.[4] The first was the realization that all cells carry a sugar coat consisting of glycoproteins and glycolipids,[5] and the second was the discovery of lectin on the cell surface of animal tissues. Lectins are a class of proteins which combine with carbohydrates both selectively and reversibly. Surface carbohydrates and lectins change in composition and population with the physiological and pathological state of the cell. Two of the best understood recognition processes involving carbohydrates and lectins are that of microbial adhesion to host cells and the adhesion of the white blood cell to blood vessels.

To cause disease, viruses, bacteria, and protozoa must be able to adhere to their host cell. Infectious agents lacking this ability are swept away from potential sites of infection by the body's cleansing mechanisms. An example of specific binding to the host cell is displayed by the sexually transmitted disease *Nisseria gonorrhoea*. Humans are the exclusive host of *Nisseria gonorrhoea* as it adheres only to human cells of the genital and oral epithelium. This mechanism of interaction introduces two possibilities for blocking the infection pathway, both demonstrated to be effective in mice. Strategies involve the saturation of the bacteria lectin by administration of its complementary carbohydrate, and the saturation of cellular carbohydrates by administration of specific lectin proteins.

Carbohydrate lectin interactions are furthering the understanding of the inflammation response. White blood cells circulate in the blood, but perform their infection fighting duties in the extravascular spaces. On tissue injury cytokines such as interlukin-I and tumor necrosis factor are released. The cytokines stimulate the venule endothelial cells to express P- and E-selectins. These selectins adhere to the presented carbohydrate coat (related to the blood group oligosaccharides) of passing white blood cells. Once attached to the wall of a venule the white blood cell can squeeze between adjacent endothelial cells and enter the extravascular space. P-selectin is stored within endothelial cells and is released in minutes, while E-selectin requires synthesis, and is released in maximal concentrations after four hours. Herein lies a mechanism for the recruitment of different white blood cells at different stages of the inflammation process. In addition the carbohydrate recognized by E-selectin is expressed on the cells of diverse tumor lines. It is thought that this binding may be integral to the metastatic process.

An understanding of carbohydrate form and function will therefore have practical applications to the prevention and treatment of a variety of ailments including bacterial, protozoal, and viral infections; inflammatory diseases such as arthritis; reperfusion injury common after cardiac illness; cancer, and in the selective delivery of drugs to specific cell lines.

B. The Language of Carbohydrate Specificity

A method of decoding the information stored in complex carbohydrates is required for the specificity of the interactions implicated above. The lock and key model formulated in 1897 by Emil Fischer proposes that molecules recognize each other by pairs of complementary structures on their surfaces, in a similar manner to a key fitting into a lock. It is thought that monosaccharides serve as letters in a language of biological specificity, words are spelled out by variations in the monosaccharide letters, by the differences in the glycosidic linkages between them and by the presence or absence of branches.

In this scenario carbohydrates would appear to present a biological language far more diverse than either proteins or DNA. Carbohydrates contain many more unique monosaccharides than DNA (4 bases) and proteins (21 principal amino acids). In addition, DNA and proteins linking points are fixed, whereas carbohydrates can link between all carbons, and in a nonlinear manner. They therefore appear to have the potential to carry more information than either DNA or proteins.

A brief overview of the nature of interactions between lectins and carbohydrates gives insight into important carbohydrate informational subdivisions. Examination of lectin binding sites indicates the possible number of sugar residues involved in these interactions. Plant lectins appear to bind to mono- and disaccharide units. Cell surface oligosaccharides can be anywhere between one and at least 20 residues. All known lectins fall into the size range of typical globular proteins forming an upper limit of the binding site of about six monosaccharides. A specific pentasaccharide sequence of heparin interacts with ATIII, although heparin segments of twenty residues produce optimal activity. It is therefore likely that carbohydrate lectin interactions generally involve less than six saccharide residues.

To understand the language of lectin biological specificity an understanding of carbohydrate three dimensional structure is required. Central to the language of carbohydrate code is the question of whether the glycosidic linkage is *rigid or flexible*. If glycosidic linkages are rigid, saccharide specificity will be dominated by the relative orientations of the monosaccharides and the resultant orientation of specific pendant groups. However if carbohydrates are flexible understanding observed specificity becomes more difficult. Many unique saccharide linkages will traverse very similar volumes of $\phi\psi$ space, on these grounds making them conformationally indistinguishable by $\phi\psi$ subdivision. In addition, the problem of relating structure to function is complicated, because the single or family of conformation(s) responsible for an interaction need not be the most stable. Therefore, when studying the molecular recognition of flexible systems, all accessible conformations should be considered, irrespective of energetic considerations in the initial analysis. The definition of an occupancy volume or parking zone may provide useful structural information.[6] This approach makes the assumption that the molecule must be parked within this volume upon receptor binding. However, it is possible that initial interactions with receptor molecules will alter conformational properties to the extent where no conformation found in solution is responsible for activity. This phenomenon has been observed in several biological systems, and has resulted in the "Rusting" of the Lock and Key Model.[7] Such examples confirm the reasonable expectation that flexible molecules distort to form optimal interactions with binding partners. A practical consequence is the frustration that will often accompany attempts to design drugs by analogy to the structures of flexible, unbound, active substances. Furthermore, during binding, flexibility must be frozen out. This implies that interactions must have favorable enthalpy, since a reduction in flexibility reduces entropic favor.

Observations of flexibility displayed by a variety of glycosidic linkages under study has led us to the hypothesis that *information, additional to the primary structure can be contained in additional interactions of side chains possibly far removed in sequence. Such interactions could localize flexible linkages to localized regions in $\phi\psi$ space, forming unique three dimensional structures. These unique conformational pockets and shapes could then be responsible for many of the interactions of carbohydrates.*

II. MOLECULAR MODELING OF CARBOHYDRATES

Due to the number of degrees of freedom in a simple disaccharide ($\approx$ 195) approximations, simplifications, and assumptions are required to enable their modeling. Past methodology has been limited since:

- Modeling studies have been dependent upon initial conditions, and the approximations, simplifications, and assumptions of each modeling approach.
- In solution, few accurately measurable conformationally dependent properties are available.
- Relationships between NMR (particularly nOe) data and molecular parameters have been imprecise.

The above problems have meant that the evaluation of molecular modeling predictions of the conformation and dynamic nature of carbohydrates in solution has been inconclusive. *Subsequently, there is an ongoing debate, reviewed in this chapter, both as to the most accurate force field for the depiction of carbohydrate solution behavior, and the nature of this solution behavior.*

In the investigation of carbohydrate conformation it is useful to subdivide the analysis into two structural partitions: (1) the preferred monosaccharide ring conformation(s), and (2) the relative orientations of the constituent pendant groups. For analysis of disaccharides and larger polysaccharides a third category is important, namely (3) the relative orientations of the monosaccharide rings.

Pyranose ring conformations are puckered so as to minimize the energy contribution due to the deviation of each ring bond angle from the ideal of $109°28'$. This results in the formation of a possible thirty eight different ring conformations, two chairs, six boats, six twist-boats, twelve half-boats, and twelve half-chairs (Figure 1). The two poles represent the energy wells of the chair conformations $^1C_4$, $^4C_1$. In unsubstituted cyclohexane the two chair conformations are at least 3 kcal mol^{-1} lower in energy than any other ring conformation, and hence are the prominent species. Substitutions of hetero atoms in the ring such as in the pyranose ring of carbohydrates and addition of hydroxyls or other exocyclic substituents further stabilize or destabilize ring conformers in relation to cyclohexane. The stable conformations of cyclohexane are in

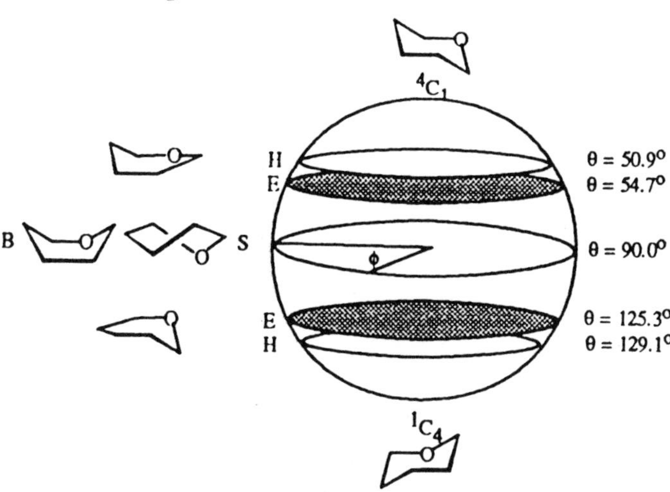

Figure 1. Graphical representation of the nomenclature for carbohydrate pyranose conformations. Chair (C), Envelope (E), and Half-Boats (HB) have one atom out of the plane. Skewed-Boat (S) conformations have two atoms on opposite sides of the plane separated by one atom. Half-Chairs (HC) have two adjacent atoms on opposite sides of the mean plane. In addition, similar ring conformations are distinguished by superscript and subscript atom numbers which correspond to the atoms above and below the mean plane, respectively. For example, ^2S_O, refers to the twist boat which has atom C2 above and the oxygen atom below the mean plane defined by the remaining atoms C1, C3, C4 and C5. Angles are representative of Cremer-Pople ring puckering parameters.[9]

relatively deep wells, even among the less stable equatorial twist boat forms pseudo rotation is somewhat hindered.[8]

A. The Anomeric Effect

When discussing carbohydrates the *anomeric effect* describes the axial preference for an electronegative substituent of the pyranose ring adjacent to the ring oxygen.

The magnitude of the anomeric effect varies with a number of factors, the most important being the identity of the electronegative group, the nature of the solvent, and the location of the substituents at other than

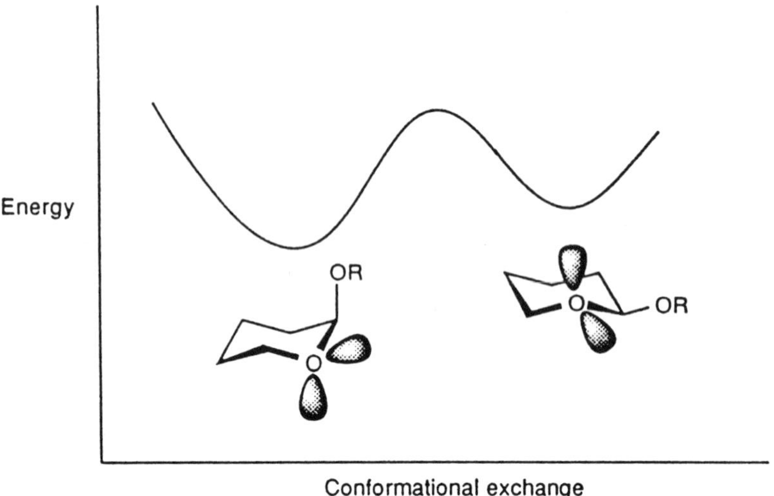

Figure 2. The anomeric effect describes the axial preference for an electronegative substituent of the pyranose ring adjacent to the ring oxygen.

the anomeric carbon. Examination of X-ray crystallographic data reveals a general shortening of the $C1-O_{ring}$ bond, a lengthening of the C–OX bond, and an increase in the $O_{ring}-C1-OX$ bond angle in the axial configurations.[10] Molecular orbital theory accounts for these observations. *The axial orientation of the electronegative substituent positions a lone pair of the ring oxygen antiperiplanar to the adjacent C–O bond and hence allows a favorable HOMO LUMO interaction* ($\approx$1.4 kcal mol^{-1}, Figure 2). This increase in bonding between C1 and O_{ring} will result in a shortening of this bond and a corresponding decrease in bonding and increase in length between C1 and OX. The increased double bond nature between O_{ring} and C1 gives C1 increased sp^2 character and therefore increases the $O_{ring}-C1-OX$ bond angle.

The *exo-anomeric* effect influences glycosidic C–O linkages in pyranosides and is therefore important in determining the relative orientations of saccharide units in carbohydrate chains. The exo-anomeric effect is a balance between electronic and steric effects. The electronic interactions in the exo-anomeric effect are similar in origin to those in the anomeric effect, except the orbitals of the glycosidic oxygen atom are

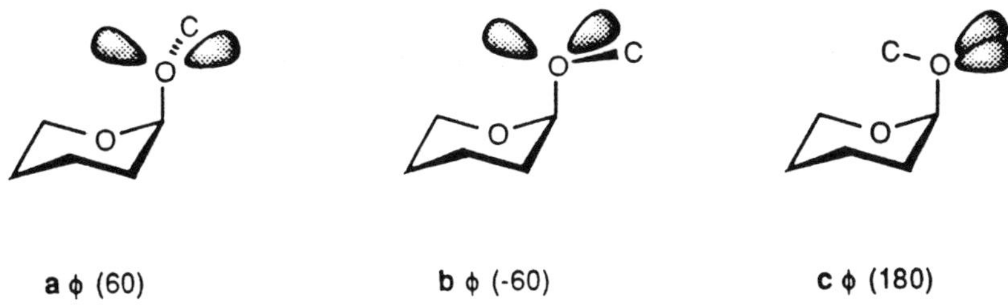

a ϕ (60)   **b** ϕ (-60)   **c** ϕ (180)

Figure 3. The three generalized gauche arrangements of groups around an axial C1–O1 glysodic bond.

involved as opposed to the O_{ring} orbitals. The exo-anomeric effect is demonstrated in Figure 3, by an axial glycosidic linkage. Conformations *a* and *c* orientate the lone pair of the exo-oxygen antiperiplanar to the adjacent C1–O_{ring} bond, hence exhibiting a favorable HOMO LUMO interaction. Therefore electronic interactions favor $a \approx c > b$. Steric interactions favor $a \approx b > c$, due to the steric effects of O_{ring}, C2, and the aglyconic group, that is, the group attached to the glycosidic linkage.

It is known from crystallographic studies, NMR measurements, and theoretical calculations that the conformational equilibrium about the C5–C6 bond in aldopyranoses depends significantly upon the configuration at C4. Figure 4 displays the definition of the orientations of the C5–C6 torsions. For the 'gluco' configuration (O-4 equatorial) the tg conformation is high in energy and the remaining two conformations are almost equally populated. For the 'galacto' configuration (O-4 axial) there is more discrepancy in the interpretation of the experimental results. However, it is agreed that all three conformers are more uniformly distributed.

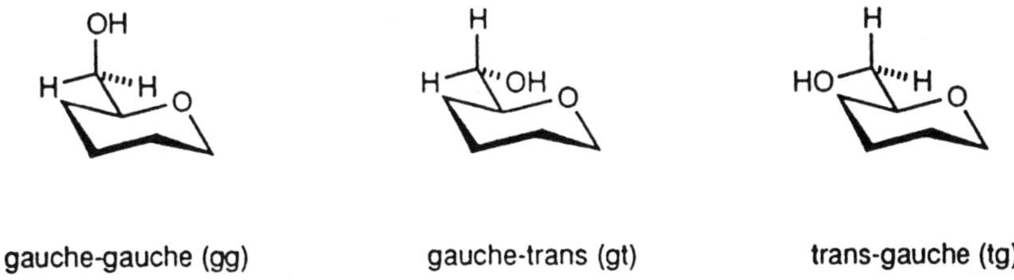

gauche-gauche (gg)   gauche-trans (gt)   trans-gauche (tg)

Figure 4. Definitions of glucopyranose C5–C6 orientations.

B. Force Fields for Carbohydrates

Modeling of the anomeric effect requires modifications to the molecular mechanical potential energy function form or parameterization. Force fields that have been modified for carbohydrate modeling are summarized below.

Hard Sphere (HS) calculations model the potential energy of a carbohydrate system as a function of the nonbonded interactions of rigid residues, expressed as a Kitaigorodsky function. The parameter set of Venkatachalam and Ramachandran[11] is used with additional parameters for the anomeric effect being calculated from ab initio calculations on dimethoxymethane. The *Hard Sphere Exo-Anomeric Potential*[12,13] (HSEA) has an additional torsional potential about one of the glycosidic angles to account for the exo-anomeric effect.

The *Potential Energy Function* (PEF) force fields developed by Rasmussen[14] have been adapted for carbohydrate modeling.[15] PEF has a simple all-atom-potential form, with energy contributions from bonded, nonbonded, angle, and single bond terms.

The *CHARMM*[16,17] potential and parameter set has been revised for application to oligosaccharides by least squares fitting of calculated vibrational and structural properties to experimentally observed values for α-D-glucopyranose.[18]

Due to the importance of glycoproteins in nature it has been desirable to modify the AMBER[19,20,21] force field to facilitate the study of these compounds. With this in mind, Homans[22] modified AMBER to ab initio $6\text{-}31G^*$ basis set calculations of several monosaccharides. Calculations were compared to the crystal structure of O-α-D-mannopyranosyl-(1-3)-O-β-D-mannopyranosyl-(1-4)-2-acetamido-2-deoxy-α-D-glucopyranose.[23] The following glycosidic torsional terms were included in the potential energy functional form.

$$E_{(\alpha_{\text{glycosides}})} = +2.15(1 + \cos(\theta - 60)) + 1.75(1 + \cos(2\theta - 60))$$

$$+ 0.85(1 + \cos 3\theta);$$

$$E_{(\beta_{\text{glycosides}})} = -1.05(1 + \cos\theta) + 1.25(1 + \cos 2\theta) + 1.40(1 + \cos 3\theta)$$

1

Homans adaptations to AMBER have been included in AMBER[*24,25] of MacroModel v3.5X,[26] as has Kollman's new 6,12 Lennard Jones hydrogen bonding potential.

Jeffrey and Taylor (MM2CARB)[27] modified MM1 and MM2 to include acetal segment parameters, by paramitization to eighteen methyl pyranoside crystal structures following ab initio calculations. This all-atom representation has been shown to reproduce ab initio energy and bond length and angle variation of several monosaccharides,[28] and produce the most accurate theoretical nOe data for O-α-D-mannopyranosyl-(α-1-3)-O-D-mannopyranosyl-β-O-Me (**1**),[29] Table 1.

To summarize, it is generally accepted that all atom force fields and minimization with respect to all degrees of freedom are required for the accurate representation of carbohydrates, as rigid residue analysis (HS, HSEA) gives only a qualitative estimate of oligosaccharide conformers.

Table 1. The nOe's Calculated by the Program DYNAMO Based upon Force Field Predictions of the Accessible Conformational Space of **1** on Irradiation of the H1 Manα Relative to the nOe on Manα-H5.

^1H	βM–$H2$	βM–$H4$	βM–$H2$	βM–$H4$
Force Field	*Rel. nOe*		*Abs. nOe*	
HSEA	1.2	2.0	1.2	2.0
HEAH	11.0	3.2	5.7	1.6
PFOS	13.0	1.7	7.8	1.0
PSOF-H	18.0	0.36	20.3	0.4
MM2(85)	8.2	1.1	7.4	1.0
MD (320 ps) PEF422	1.1	1.2	5.9	6.5
observed	1.8 (0.4)	0.7 (0.2)	1.8 (0.4)	0.7 (0.2)

Note: nOe data is at 300 MHz, $\tau_c = 1.1 \times 10^{-10}$.[29]

A degree of agreement exists between AMBER, MM2CARB and PEF relaxed $\phi\psi$ potential energy surfaces for the disaccharide **1**. *However, as the differing potential surfaces and solvent models have not been fully explored, and methods of comparison to experimental data are inadequate, comparison and experimental evaluation of the differing force fields remains inconclusive.*

III. INTEGRATING CARBOHYDRATE POTENTIAL ENERGY SURFACES

A. Contour Maps—Relative Orientations of Monosaccharides

The relative orientations of saccharide units are expressed in terms of the glycosidic linkage torsional angles phi (ϕ) and psi (ψ) which have the definition ϕ = H1-C1-O-C'X and ψ = C1-O-C'X-H'X, and are analogous to ϕ_H and ψ_H of IUPAC convention (Figure 5).

Contour maps enable graphical depiction of energy change as related to the indices of alterations in relative orientation of the monosaccharides, the glycosidic angles ($\phi\psi$). This renders them a useful tool for representation of a three dimensional cross section of the complex conformational space of disaccharides. They indicate the shape and position of minium, the routes for inter-conversion between conformers, and the heights of the transitional barriers.

There are many different methodologies for calculating contour maps. Their limitations and implications on carbohydrate conformation are

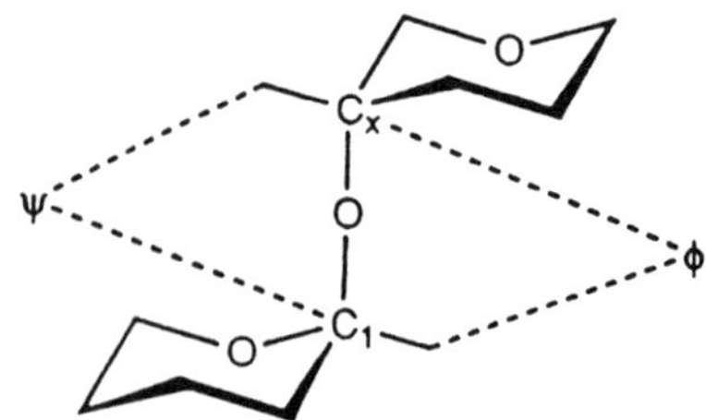

Figure 5. The relative orientations of saccharide units expressed in term of the glycosidic linkage torsional angles phi (ϕ) and psi (ψ). In this chapter the eclipsed orientation is defined as 0° and clockwise rotation as positive.

reviewed below. The extent to which early rigid residue modeling studies are responsible for the conceptualization of disaccharides tumbling as rigid bodies in solution becomes apparent by comparison of rigid and relaxed conformational maps. A comparison, Figure 6, of the rigid (a), and relaxed (b) residue contour maps for **1** shows that the relaxed map depicts

- a considerably larger accessible potential energy surface,
- a reduction of the magnitude of barriers between minium,
- lower energy minium far removed from the initial starting geometry, the methodology therefore locating previously overlooked minium.

Similar differences as between rigid and relaxed maps have been observed for α-maltose,[30–33] sucrose,[34,35] and -α-L-fucopyranosyl-(1-2)-O-β-D-galactopyranosyl-β-O-Me (2)[36] and β-cellobiose.[37,38] This pronounced effect of pendant group orientation on conformational flexibility was first demonstrated in conformational maps of acetylcholine and β-methyl acetylcholine.[39]

Rigid Residue Maps

Early contour maps are termed Ramachandran,[41] rigid residue, or hard sphere potential surfaces, as the constituent monosaccharides are assumed to be rigid, with pendant groups fixed. The dihedral driver is a molecular mechanical method allowing exploration of $\phi\psi$ space in a systematic automated fashion. $\phi\psi$ torsions are sequentially rotated in small increments through space, as depicted by Figure 7. At each point on the grid energy contributions calculated from the force field in use are evaluated.

Rigid residue maps often utilize crystal structural data as the starting conformation of constituent monosaccharide residues. In crystal structures all pendant groups are orientated in steric and electronic positions which favor the specific $\phi\psi$ orientations of that crystal conformation. As the $\phi\psi$ values are altered, steric interactions occur which are unable to relax between these pendant groups. These steric interactions cause a rapid increase in energy. This effect is especially prominent in sterically crowded molecules such as carbohydrates.[42] In addition, surveys of a large number of known crystal structures along with and supported by

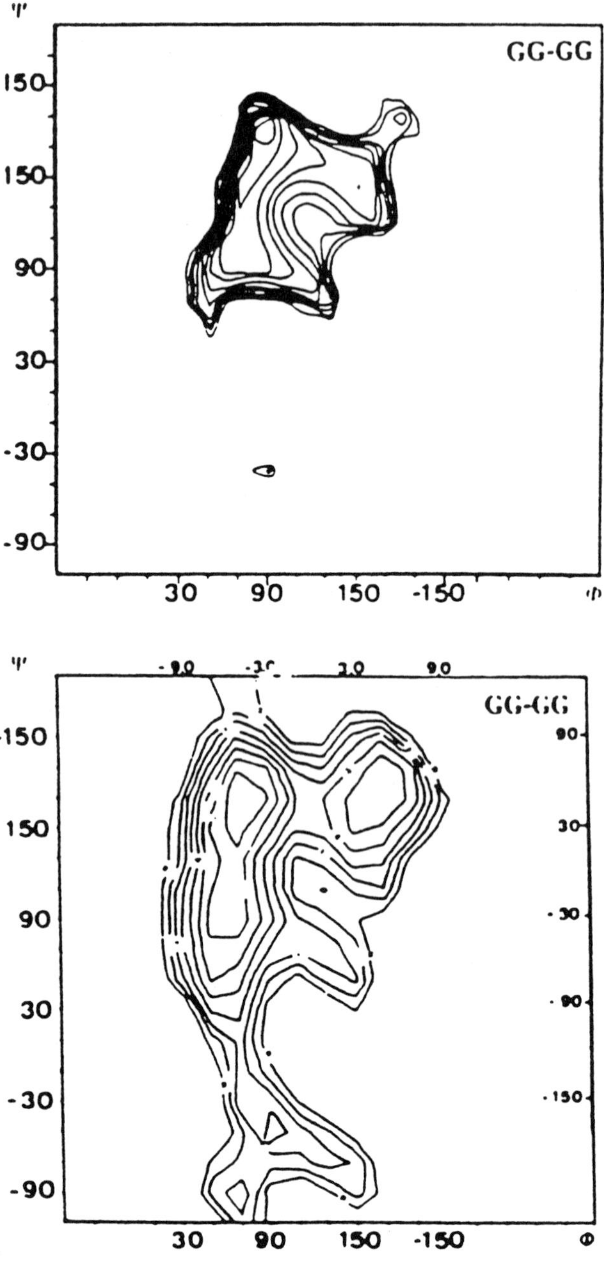
GG-GG
GG-GG

Figure 6. Rigid (top) and relaxed (bottom) contour maps of Man-α(1-3)Man-α-O-Me. Contours are drawn at 1 kcal mol^{-1} increments above the minium to a maximum of 10 kcal mol^{-1}. Hydroxymethyl groups are locked in a gauche-gauche gauche-gauche arrangement. The starting structures was taken from the crystal structure of Warin et al.[23] φ and ψ are defined by O5-C1-O1-C'3 and C1-O1-C'3-C'4, respectively. PFOS rigid potential surface (top). Energy contributions are from van der Waals interactions, torsional, exo-anomeric and hydrogen bonded potential. The ring geometry is invariant, hydroxylic hydrogen atoms are ignored. MM2CARB relaxed potential surface (bottom). 20° × 20° grid incorporating stretching, bending, stretch bend, torsional, dipolar contribution and van der Waals interactions. There is no explicit hydrogen bonding potential. Hydroxymethyl groups are fixed in a gauche-gauche gauche-gauche arrangement. Reprinted with permission.[40]

---

semi-empirical calculations reveal small but important variations in pyranoid ring geometries and orientations of pendant groups with φψ values. These are dependent upon the anomeric and exo-anomeric effects and emphasize the need for a model to include bond length and angle degrees of freedom.

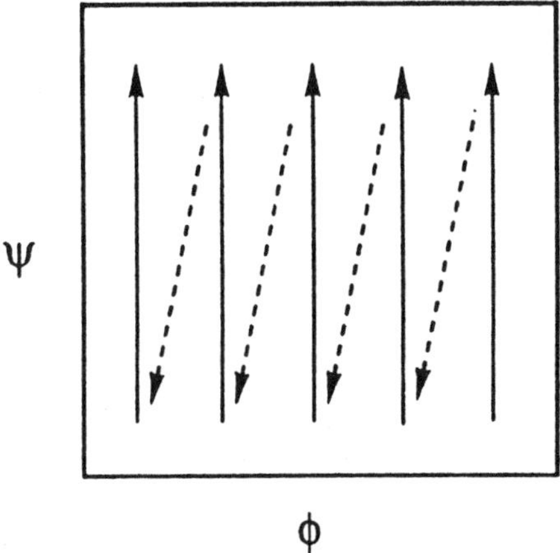

Figure 7. The dihedral driver mechanism of moving through φψ space.

Relaxed Maps

Relaxed residue maps considerably reduce the steric energy repulsions encountered in rigid residue maps. The strain produced by steric interactions inherent from rotation of monosaccharide residues is relieved by the inclusion of bond length and angle adjustment in the form of minimization, with respect to all degrees of freedom of the system, at each point in space. Production of a map with such a minimization along all degrees of freedom (except $\phi\psi$) is clearly a much larger computational task than that involved in generating fixed geometry maps. A penalty energy function is utilized to constrain $\phi\psi$ to the angles of interest, thus allowing minimization of all other degrees of freedom. The penalty energy term is identical to the torsional term of the force field with a large force constant. The energy contribution from this term is not included in the molecular energy but its contributions to the derivative and the second derivative of the energy potential, serve to confine the torsion of interest to the chosen value.

During minimization pendant groups move to the nearest minimum downhill of the starting point. In the process of driving the molecule through unfavorable regions of $\phi\psi$ space with the dihedral driver mechanism, large steric interactions can sometimes cause pendant groups to overcome torsional activation barriers. This results in minimization to a different local well. Due to the dihedral driver mechanism this flexibility is irreversible, and cumulative, as the previous structure is used as the starting geometry for the next grid point. The presence of flexibility of this nature became apparent in relaxed maps with different relative energies at 180° and –180°.[38] This flexibility renders potential surfaces created in this way conformationally misleading as it is not known which conformation one is looking at. Subsequent developments in methodology to avoid the passing on of these inelastic deformities are summarized below.

Each point in $\phi\psi$ space can be rotated to from the original monosaccharide geometry. This however recreates the problem of minimization failure for which the original dihedral driver was created. Initially the monosaccharide geometries have bond angles close to ideal values and hence the cubic bend term in the force field is not deactivated, but as the minimization proceeds bond angles are occasionally severely deformed due to large steric interactions. This results in huge contributions from the cubic term, and the blowing apart of the molecule. More realistic energies for these unconverged points can be calculated by:

- extrapolation of the energy from a nearby point (i.e., SURFER program[44]),
- using different starting geometries which avoids large unfavorable steric interactions,
- using a sufficiently optimized geometry from an adjacent point in $\phi\psi$ space as a starting geometry,
- adjusting the C–O glycosidic bond angle to 124° to allow groups time to move away during minimization while the bond angle reverts back to 109°.[45]

The above methodologies are very time consuming, as each point which results in a unrealistic energy requires human intervention. More temporally efficient methods are mentioned below:

- local relaxed maps explore $\phi\psi$ space in small increments radially outward from the global minium.[46]
- All pendant groups can be localized to the appropriate low energy well by a flat bottomed restraining potential.[36] This allows unaffected minimization of all pendant and ring degrees of freedom to their nearest low energy minima, and simultaneously disallows any transitions of these degrees of freedom to other low energy wells.

This method affords the complete automation of relaxed map generation, the process taking <3 hours on a IBM RISC 320h (18×18 grid). It is useful to calculate several different maps for each disaccharide each with different orientations of pendant groups especially the C5–C6 torsions. We believe multiple relaxed maps reveal more $\phi\psi$ conformational information than adiabatic maps as the effect of pendant group orientation on conformational space can be evaluated.[36]

Adiabatic Maps

The differences between rigid relaxed and adiabatic conformational maps can be understood by comparison of the potential energy surface of the system and the corresponding potential energy surface represented by the modeling method. Rigid residue maps represent a two dimensional cross section, (the independent $\phi\psi$), of a $3N - 6$ dimensional surface where N is the number of atoms. Relaxed maps allow minimization of the internal coordinates of bond lengths, and bond and torsional angles to local low energy wells. Thus relaxed maps represent a larger cross sectional window of a given potential energy surface than rigid residue maps. However, as minimization will only lead to conformations 'down-hill' from the starting structure, the torsional dimension where most conformational variation occurs is limited to only one orientational well. It is possible that rotation of pendant groups over torsional barriers could produce lower energy conformations at that point in $\phi\psi$ space. By way of example, sucrose has a $\phi\psi$ low energy conformation that is stabilized by an interglycosidic hydrogen bond, which in turn depends upon pendant group orientation.[34] Ideally at each point in $\phi\psi$ space investigation of all possible combinations of pendant group orientations is required (i.e., $3n$ different conformations at each point in $\phi\psi$ space, where n is the number of pendant torsions). This results in 3^{12} (531 441) conformations for a simple disaccharide such as Man-α-(1-3)Man-β-OMe (**1**) and 319 (1.16×10^9) conformations for a more complex disaccharide typical of heparin.

Adiabatic maps attempt to represent the lowest energy of all possible pendant group orientations at each point in $\phi\psi$ space. The effect on the potential energy surface is demonstrated by the adiabatic map of Man-α-(1-3)Man-α-O-Me, (**1**) (Figure 8). This map is calculated from nine relaxed maps each originating from a different low energy conformations, introducing an additional degree of freedom for each starting structure used. On comparison with the corresponding relaxed maps (Figure 6), adiabatic maps are flatter, allow greater freedom about the glycosidic bonds, locate additional minimum (e,f), and reduce the barriers between the minimum.

At present there are several different methods for calculating adiabatic conformational maps:

1.   The energy at each point in $\phi\psi$ space for several different starting geometries is evaluated. The lowest energy for each point is used to generate the map. An extra dimension of conformational flexibility is incorporated for each starting geometry used. With the appropriate choice of starting structures much of conformational space can be

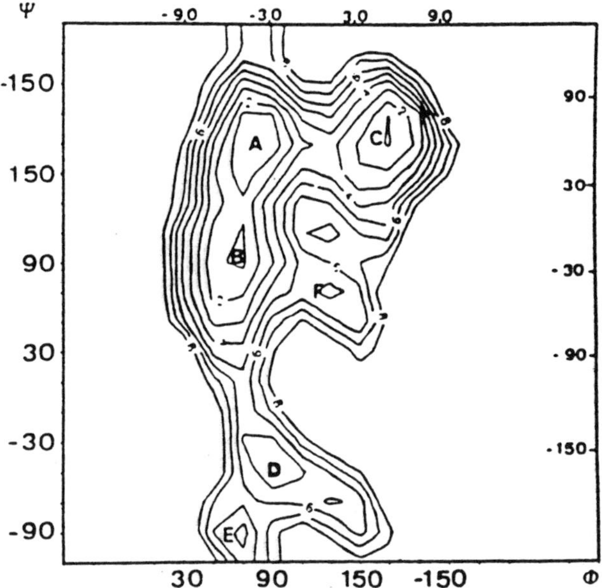

Figure 8. The MM2CARB adiabatic relaxed potential energy surface of Man-α(1-3)Man-α-O-Me. Each point represents the lowest energy conformation from nine relaxed maps. Each relaxed map was calculated from a starting structure with different low energy orientations of the hydroxymethyl groups. Iso-energy contours are drawn at 1 kcal mol^{-1} intervals above the minium. Reprinted with permission.[40]

represented. However as previously mentioned we believe that more conformational information is contained within the relaxed maps before their superimposition.

2.   Random Molecular Mechanics (RAMM)[47] grid searches the orientation of pendant groups at each point in ϕψ space. At each point in ϕψ space, 1000 steps of a random walk procedure varies pendant group orientation and evaluates unrelaxed energies. Only the resultant lowest energy structure is optimized, and accepted as the (MM2CARB) energy for that point in ϕψ space. This methodology avoids problematic combinations of pendant orientations, but does not guarantee that the structure at each ϕψ point is the lowest energy combination.

3.   Prudent assent moves through ϕψ space in a way which is dependent on previous minimizations. Large steric interactions are minimized

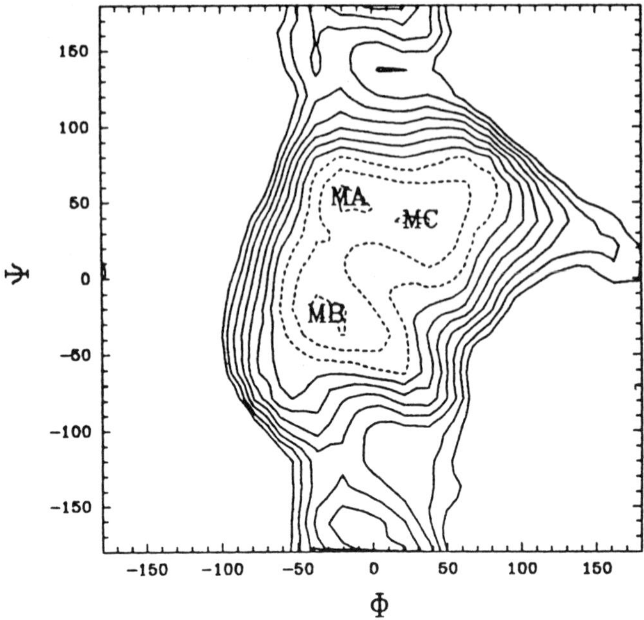

Figure 9. MM2* GB/SA[26] relaxed map of **1**, from the lowest energy conformation located (C5–C6 gt-gt) after a fully converged conformational search. Iso-energy contours are drawn at 1 kcal mol^{-1} intervals above the global minium. Space below 3 kcal mol^{-1} is denoted by dotted lines.[36]

by doing the most favorable geometries first, in a way similar to the local relaxed map. It makes use of inelastic deformations that decrease the energy by recalculating the energies of surrounding geometries using the new lower energy structure as the starting geometry. On average the energy for each point in φψ space is calculated twice. The use of molecular mechanics in this form enables surmounting of energy barriers, and location of lower energy minium. This method on average appears to surmount ≈10 rotational barriers and results in significant energy reductions, thereby incorporating a similar number of degrees of freedom into the conformational map.

To summarize, the apparent flexibility about the glycosidic linkages of disaccharides has increased as a greater number of dimensions have been introduced into contour map modeling. Contour maps are dependent on the starting geometry used.[49] This dependence decreases in the order rigid > relaxed > local relaxed > Prudent > RAMM. The inclusion of solvation into contour maps by the GB/SA continuum model of MacroModel results in additional flexibility (Figure 9).[36]

B. Molecular Dynamics versus Mechanics and Contour Maps: A Strategy

With ongoing advances in computer power and algorithm development it is becoming possible to investigate molecular conformation and interactions in the solution phase where most bioorganic chemistry occurs and to explore fully all the bond length, bond angle, and torsional degrees of freedom in a system. Disaccharide hypersurfaces consist of >12 torsional dimensions. This demands concurrent development of methods for conceptualization and representation of these hyperspaces. To this aim we believe that the description of low energy patterns by molecular mechanic conformational search protocol and graphical representations of cross sections of the hyperspace play an important role, building a mental model of the potential surface and furthering the interpretation of the more accurate converged molecular dynamic integral. A presentation of selected demonstrative modeling results follows.

The controversy of the degree and time scale of dynamic motion in carbohydrates can be resolved by generating integrals, by fully converged conformational search, molecular dynamic simulations and relaxed $\phi\psi$ energy maps, over the MM2* and AMBER*, vacuo and solvent potential energy (GB/SA solvent model) surfaces. Such investigations have allowed accurate comparison of the different potential surfaces and approaches to evaluating the integral over these potential surfaces.

All three of the above methods of integration were found to play a useful role in the full description of conformational space. Converged conformational searches generated 366 and 83 conformations for **1** and **2**, respectively. Torsional analysis[50] of these results allowed grouping into <20 basic conformational patterns (e.g., $\phi\psi$ wells, C5–C6 orientations, hydrogen bonding arrangements, etc.), of which various combinations made up all the low energy conformations. Correlation between conformational patterns and torsional parameters allowed numeric analysis and the abstraction of conformational properties. This basic knowledge of the conformational hyperspace was invaluable for interpretation and understanding of subsequent molecular dynamic simulations. By way of example, two hydrogen bonding networks exist around the Fuc ring of **2**. The hydrogen bond F2H″···F3O occurs when F2 ≈–50, and the hydrogen bond F2O···F3H occurs when F3 ≈–50. This knowledge facilitates the interpretation of Figure 10.

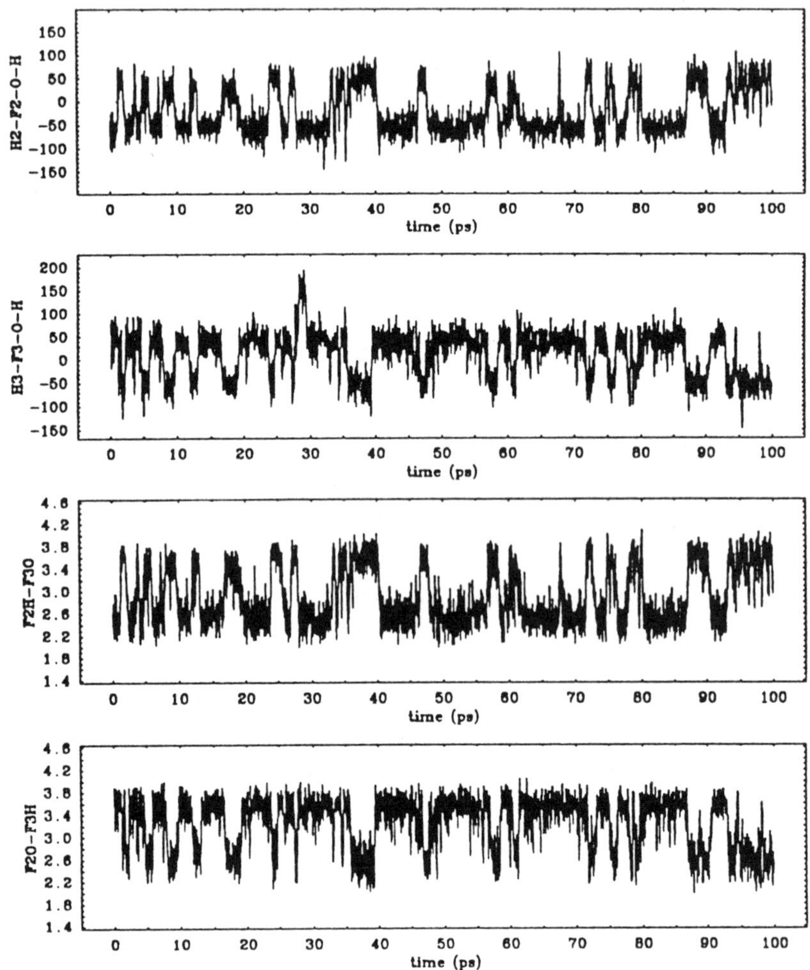

Figure 10. Torsional angle and distance monitoring of a 100ps stochastic dynamic 300 K MM2[*], GB/SA[26] simulation of **2**. The mutually exclusive hydrogen network around the Fuc ring is well demonstrated.[36]

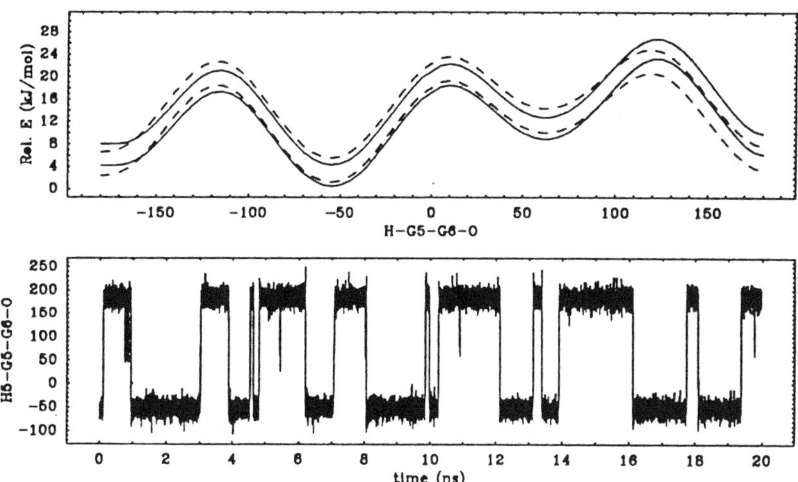

The effect of C5–C6 and hydrogen bonding pendant group patterns can be investigated by relaxed map analysis from appropriate geometries. The geometries were selected from conformational search results. Relaxed maps were generated from selected conformational patterns to evaluate the effect of the selected patterns on $\phi\psi$ space. Relaxed rotational profiles of single torsional angles in conjunction with the Eyring equation affords an estimation of the rotational periodicity, and molecular dynamic convergence times. By way of example, rotational profiles

Figure 11. MM2*, GB/SA[26] relaxed rotational profile of **2** Fuc C5–C6. Each plot relates to a different family of conformations, (top). Torsional angle monitoring of the same angle during a 20 ns stoichastic dynamic simulation at 300 K, MM2*, GB/SA[26], (bottom).[36]

of the C5–C6 pendant group of **2** approximated the barrier of rotation to be $\approx$18 kJ mol^{-1} (Figure 11, top). This corresponds to a rotational periodicity estimated by the Eyring equation to be $\approx$500 ps. Stoichastic dynamic simulations displayed a similar periodicity (Figure 11, bottom).

Significant differences in the population of several conformational groups,[25,36] have been observed between integrals of the conformational surface generated by molecular mechanic and dynamic methodology. These differences are attributed to:

- The nature (or shape) of low energy regions on the potential surface. The molecular mechanic assumptions of, conformational equivalence of enthalpic change and of entropic contribution between different conformations induced by temperature are inaccurate in flexible molecules, see next chapter.
- The relationships between low energy regions on the potential surface. Systems of correlated motion and thermodynamic independence require evaluation before accurate descriptions of flexible molecules are possible.

Our studies have shown that molecular dynamic integrals present a more accurate representation of both experimental NMR coupling and nOe data. Therefore in flexible molecules it appears it is misleading to determine room temperature solution properties by conformational search procedures, or by relaxed map analysis, although they play an integral role in the description of complex conformational hypersurfaces. *Molecular dynamics in solution therefore becomes the method of choice for integration of the potential surface, in molecules where convergence of simulations is possible.*

Fully converged molecular dynamic simulations are possible when barriers to transitions are small enough to be frequently crossed in the time of simulation. By way of example, if the highest barrier of rotation in a carbohydrate is $\approx$20 kJ/mol, this will correspond to approximate rotational periodicities of $\approx$500 ps (Figure 12). Any simulation at room temperature of duration less than the time required for statistical sampling of the slowest transition will be biased towards the starting geometry.

By way of example, many molecular dynamic integrals of conformational space in the past have been generated from multiple short trajec-

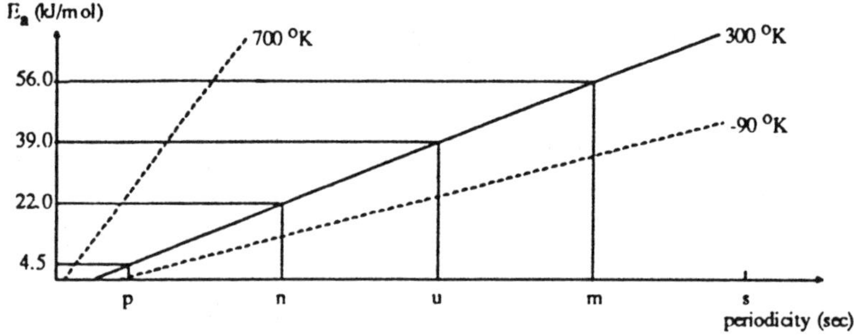

Figure 12. Graphical depiction of a rearrangement of the Eyring equation. The transmission coefficient is assumed to be 1.0.[36]

tories (<20 ps,[29] <160 ps[59]) As this period is less than many of the rotational periodicities there has been insufficient time for the initial conditions of a system to be forgotten. For this reason any integral of disaccharide conformational space generated by multiple dynamic simulations from differing starting geometries of this time scale (<500 ps) will be biased towards the population distribution of the starting geometries. The shorter the simulation lengths the more pronounced the biasing. It is therefore recommended to conduct simulations of indefinite length, only halting them if convergence is likely. Convergence is likely when: torsional distributions are the same for different simulations, all accessible areas of conformational space identified by molecular mechanic minima have been traversed, and simulations are many times longer than the longest estimated rotational periodicity (Eyring equation).

IV. FLEXIBILITY VERSUS RIGIDITY

To illustrate the position of current research in modeling carbohydrates we review three selected groups of molecules. Blood group oligosaccharides and asparagine transaminases have been central to the glycosidic flexibility versus rigidity dilemma. The third group demonstrates additional flexibility, that of the iduronic acid ring in heparin. The iduronic acid example of ring flexibility demonstrates the benefits of integrating molecular modeling and NMR (coupling constant) data.

Up to the advent of the last decade modeling studies of carbohydrates were limited to reduced dimensional contour map analysis in vacuo. These assumptions and there sequalae are still having influence on the current conceptualization of rigid tumbling of carbohydrates in solution.[51] For a demonstration of the importance of depiction of macroscopic dynamic motions for the conceptualization of natural phenomena one does not need supercomputers and in depth modeling studies. Debate on the motions undertaken by a galloping horse, were not resolved until the advent of slow motion photography.[52] A similar dependence on dynamic motions has been demonstrated on a smaller level by Karplus and co-workers[53] who showed that dynamic fluctuations are required to open a transition pathway to enable the binding of oxygen to hemoglobin and myoglobin.

The last few years has seen the emergence of molecular dynamic investigations of carbohydrate systems. Bush et al., in a series of papers studying blood group oligosaccharides concluded that glycosides were rigid. Initial conclusions by Carver et al., that N-linked glycopeptides adopt well defined single conformations, were discarded under closer scrutiny, and conformational averaging was invoked in attempts to explain experimental nOe data, particularly of the Manα(1-6)Man linkages. Several papers by Edge et al. and Homans et al. support rigidity of Manα(1-2)Man and Manβ(1-3)Man linkages. Bush et al. propose that differences observed between their own and the Carver et al. studies indicate real differences between the blood group 'flexible' and N-linked 'rigid' saccharides.

A. Blood Group Oligosaccharides

Blood group types are determined by the non-reducing antigenic terminals of blood group A, H and Lewis blood group determinants found on mucus type glycoproteins (Figure 13). They are attached to the peptide by O-glycosidic linkages between α-GalNAc and serine or tyronine residues. Our own studies have focused on the disaccharide D-α-L-Fucopyranosyl-(1-2)-O-β-D-Galactopyranosyl-β-O-Me (2) present in a majority of the blood group determinants (Figure 13).

Pioneering Hard Sphere calculations on the trisaccharide α-L-Fuc-(1-2)-α-D-Gal-(1-3)-β-D-Gal-OR located two low energy wells for the Fuc-(1-2)-α-D-Gal segment (ϕ,ψ) (55,20) and (50,10) and a single minimum on inclusion of the anomeric effect (HSEA) (55,20). Leumeaux proposed that blood group oligosaccharides have relatively rigid conformations that are primarily determined by van der Waals interactions and

Oligosaccharide	Primary Structure

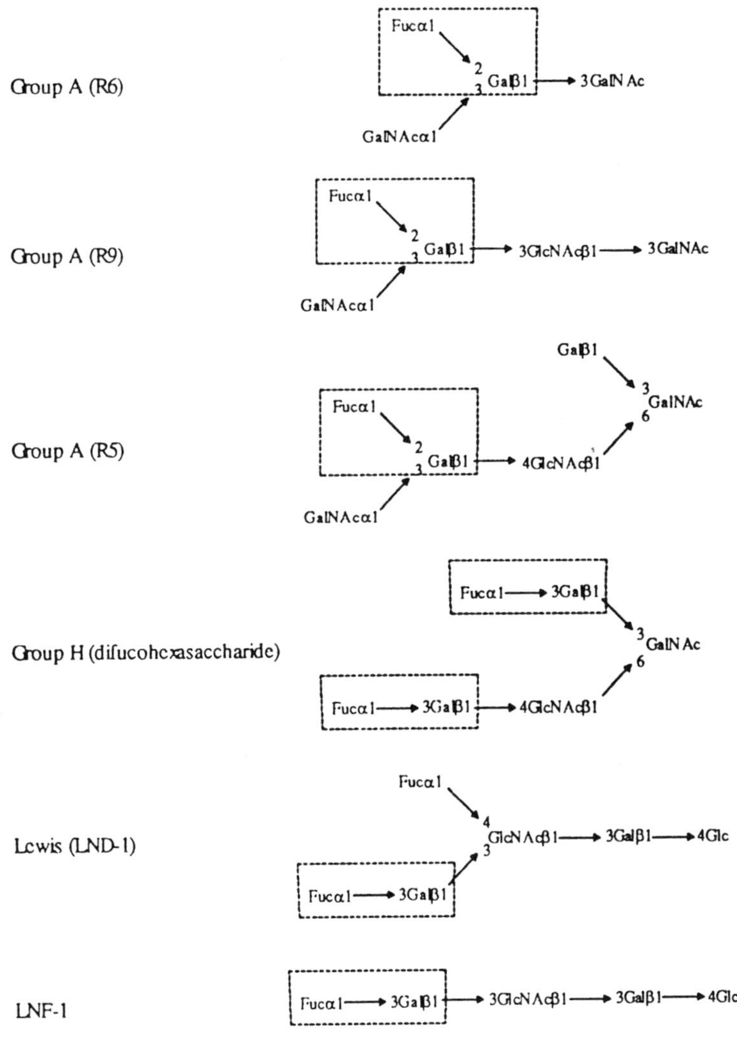

Figure 13. Schematic representation of the non-reducing terminal residues of group A, H and Lewis blood group determinants. **2** is represented by the boxed region.

torsional potentials. Observations of limited alterations of relative nOe of **2** with varying temperature have lead to the assumption that **2** tumbles as a rigid body in solution. It was concluded that the major determinants of conformational behavior must therefore be nonbonded interactions.[55] Calculation of a rigid residue map[56] yielded a steeply rising potential surface. This was taken to support the predictions of rigid tumbling, as a change in temperature from 0–100°C would not be expected to cause a change in oligosaccharide conformation as the potential energy barriers between minium MA and MB $\gg kT$.

Molecular dynamic trajectories investigating a cumulated total of 400 ps of real time (all trajectories < 160 ps) using the CHARMM program, in conjunction with the all atom force field PEF 422, demonstrated glycosidic and C5–C6 torsional transitions with periodicity of ≈50 and 30 ps, respectively. However, as all glycosidic transitions to MB quickly returned to the global minimum well MA, and as rates of transition were decreased by the incorporation of an additional Gal unit, it was concluded that the Fuc-2-β-Gal linkage is restricted to MA, tumbling rigidly in solution.[57] These conclusions have been supported by a series of nOesy and molecular modeling studies assuming isotropic tumbling and no conformational averaging.[1,58–60] NMR studies of **2**[61] combined with calculations, assuming isotropic tumbling ($\tau_c = 5.6 \times 10^{-11}$), and neglecting conformational averaging and internal motion) related the nOe between H1 Fuc and H2 Gal to a internuclei distance of ≈2.1 Å. It was concluded that the O-glycoside bond in **2** is relatively rigid in aqueous solution.

We dispute these conclusions[36] and have found that the dependence of the FH1-G2H nOe on $\phi\psi$ is similar for both MA and MB (Figure 17), which we confirm to undergo fast conformational exchange at a rate similar to that observed by Bush et al. We propose that nOe data does not vary dramatically with temperature because MA and MB are already undergoing exchange at a rate extremely fast on the NMR time scale, and entropy effects (e.g., temperature) do not significantly affect population distributions.

B. Asparagine Transaminases

An integral feature of many cell surface and secreted proteins is the presence of one or more oligosaccharide units covalently linked to

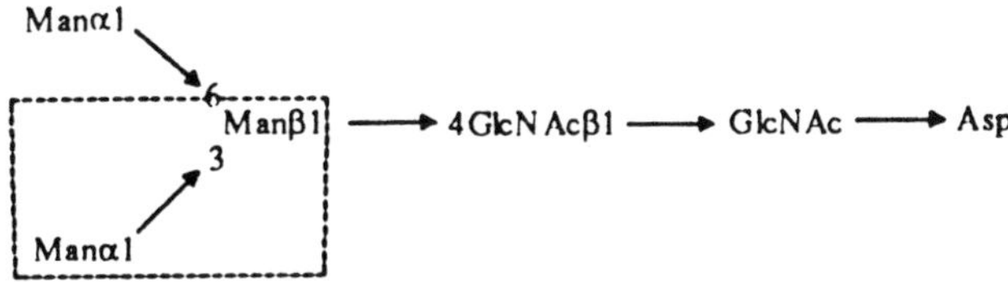

Figure 14. Schematic representation of the pentasaccharide core common to N-linked glycoproteins. Substitution at the Manβ residue of the core gives rise to the α(1-6) and/or α(1-3) antenna. Antenna include both linear sequences and further 3,6 branch points. **1** is represented by the boxed region.

asparagine. Their function is generally unknown despite intensive investigation. They are proposed to have important roles in intracellular migration, secretion of glycoproteins, biological activity of glycohormones, regulation of protein turnover, and activity and metastasis of tumor cells.[62]

In a series of papers Carver et al.,[29,63–71] coupling both NMR techniques and molecular modeling studies, have attempted to determine the solution conformation properties of α(1-3), α(1-2), β(1-2), β(1-4) and α(1-6) glycosidic linkages in a series of synthetic oligomannosides which are analogues of the tri-mannosyl core common to N-linked glycoproteins (Figure 14). Modeling studies have placed special emphasis upon the disaccharide **1**.

Agreement has been found between experimental nOe data, and theoretical nOe data calculated from crystal structure, rigid residue modeling calculations (Figure 9), and rigid map investigation. This, coupled with a 10ps molecular dynamics simulations using the GROMOS force field in which no transitions were observed[72] lead to the initial perceptions of the rigidity of both α and β (1-2), (1-3), (1-4) glycosidic linkages. Static, reduced dimension, molecular modeling studies have focused upon φψ contour maps. Calculations progressed from rigid residue maps, using a variety of force fields, hard sphere (HS), exo-anomeric effect (HSEA), hydrogen bonding (PFOS), to relaxed adiabatic potential maps of all atom force fields (MM2). Concurrent to this progression from rigid to relaxed residue maps predictions of a disaccharide conformation tumbling rigidly in solution and consistent with crystal and NMR geometries

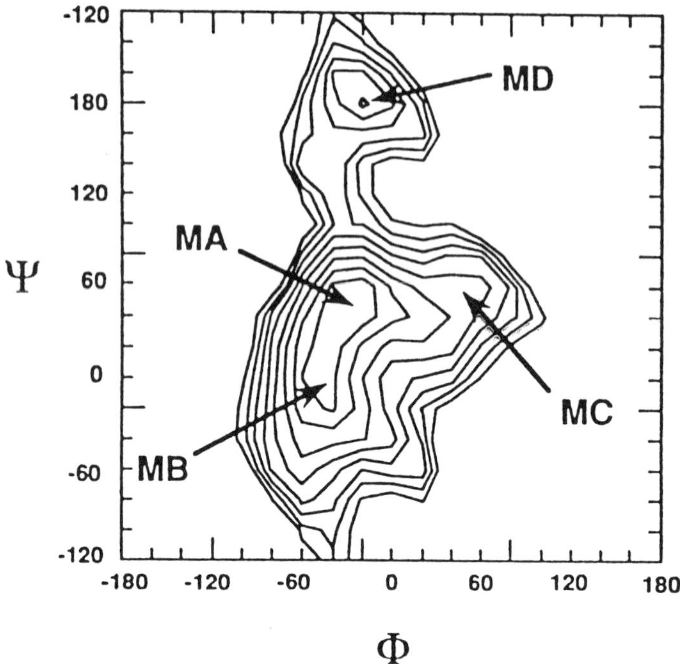

Figure 15. MM2(85) adiabatic relaxed potential energy surface for **1**, calculated by the method of Imberty et al.[40] C5–C6 {gt-gt, gt-gg, gg-gt, gg-gg} rotomers were considered. Reprinted with permission.[29]

have become less dogmatic, to the point where the possibility of average or virtual conformations attracted consideration.[67]

Conformational averaging and its effect on the nOe is discussed at a later point in this review. It is sufficient to say at this point that Carver et al. were the first to include conformational averaging in the comparison of theoretical and experimental nOe data in carbohydrates. A brief summary of this methodology follows. The statistical mechanical formulation DYNAMO,[67] incorporates conformational averaging, into the calculation of intraresidue ^1H–^1H nOe data. Conformational information is integrated over $\phi\psi$ contour maps. This affords a more rigorous test of predictions of accessible conformational space. Conformational maps were subdivided into a grid of 3° (rigid) or 20° (relaxed) resolution. The relative population of each state is calculated by transforming the steric energies E($\phi\psi$)i into a Boltzmann distribution, where the population *P* of state *i* is given by equations 8 and 9 (next chapter). Ensemble averaged cross relaxation rates, for each state (i) taking all ^{1}H–^{1}H vectors (I ≠ S) into account, were calculated from ensemble averaged values of r_{IS}^{-6}. Oligosaccharides were assumed to tumble isotropically in solution, with no internal motion, (**2** $\tau_c = 1.1 \times 10^{-10}$). Assemble averaged cross and spin lattice relaxation rates were used to solve the series of simultaneous equations defined by Noggel and Schirmer and the resultant nOe were compared to experimental values.

As the molecular models progressively incorporated more degrees of freedom agreement between the predicted and experimental inter-residue nOe between H-1 αMan and H-2 βMan increased (Table 1). The most accurate reproduction of nOe data came from the MM2(85) relaxed potential energy surface.[29]

Dynamic simulations of **1** in vacuo with Rasmussen's all atom force field PEF422 were contained in the 8 kcal mol^{-1} contour of the MM2(85) map.[29] Several transitions between MA and MB were observed. Coordinates were recorded every 10 fs and running averages of all ^{1}H–^{1}H r_{IS}^{-6} every ps were used to calculate the ensemble average nOe's (DYNAMO). Agreement with relative nOe experimental data is reasonable, Table 1. Conformational space was sampled by 16 × 20 ps trajectories from ten different starting geometries, four from each minimum, resulting in a total of 320 ps. Dynamic results where in less agreement than

Table 1. The nOe's Calculated by the Program DYNAMO Based Upon Force Field Predictions of the Accessible Conformational Space of **1** on Irradiation of the H1 Manα Relative to the nOe on Manα-H5[a]

^{1}H	βM–H2	βM–H4	βM–H2	βM–H4
Force Field		*Rel. nOe*		*Abs. nOe*
HSEA	1.2	2.0	1.2	2.0
HEAH	11.0	3.2	5.7	1.6
PFOS	13.0	1.7	7.8	1.0
PSOF-H	18.0	0.36	20.3	0.4
MM2(85)	8.2	1.1	7.4	1.0
MD (320 ps) PEF422	1.1	1.2	5.9	6.5
observed	1.8 (0.4)	0.7 (0.2)	1.8 (0.4)	0.7 (0.2)

Note: [a]nOe data is at 300 MHz, $\tau_c = 1.1 \times 10^{-10}$.[29]

the static picture, because $\phi\psi$ conformational space was biased towards the starting geometries MD and MC, as **1** has glycosidic rotational rates $\tau_{im} > 20$ ps. Carver et al. conclude that their NMR nOe data and molecular modeling results are not consistent with a single conformation for the mannose(1-3)oligosaccharide glycosidic linkages.

A 500 ps simulation of the Manα(1-2)Manα glycosidic linkage using AMBER revealed a dampening of molecular fluctuations with explicit water treatment. ^{1}H–^{1}H distances were in good agreement with distances calculated from nOe's ($\tau_m = 250$ ms assuming linear buildup approximation and rigid isotropic tumbling) by the relaxation matrix approach. These results were considered further evidence to the conformational rigidity of carbohydrates in solution. This is in spite of common short lived (≈ 1 ps) transitions to the (ϕ,ψ) (30,–10) minimum of vacuo dynamic simulations, but not depicted in the AM1 relaxed residue map reported in the same paper.

Several of the modes of internal motion within saccharides occur on a time scale >500 ps, for example, C5–C6 rotation (Figure 11). For this reason we[36] have conducted several long (20 ns) stochastic dynamic trajectories using the MM2* force field and GB/SA continuum solvent treatment, utilizing MacroModel v3.5X,[26] of the Manα(1-3)-Manβ linkage **1**. These simulations unveiled reason for the above mentioned flexibility rigidity dilemma. One simulation displayed prolonged rigidity (20 ns) in the MA conformation. However a simulation from the lowest energy conformation in the gt-gg family (+1.70 kJ mol^{-1}) displayed multiple transitions between MA, and MB of periodicity <10 ps and virtually free rotation between MA and MC, for a period of 5 ns (Figure 16). Subsequently a flip of C5–C6 to the gt-gt orientation resulted in the localization of the simulation to MA.

To summarize the transition time away from the ManαC6O…ManβC2H hydrogen bond possible in ManαC5-C6gt is long (>20ns), the transition time towards the MA hydrogen bound form is short. (<5ns). The small energetic preference for MA over MB, and similar barriers for transition both between MA/MB and G5gg/gt/tg in both the hydrogen and non-hydrogen bonded forms, and the presumed entropic favoring of the orientation synonymous with flexibility between MA, MB and MC, appears to be in contradiction to the above results. We however believe that this is further demonstration of the important influence the 'shape' of the potential energy surface plays in determining

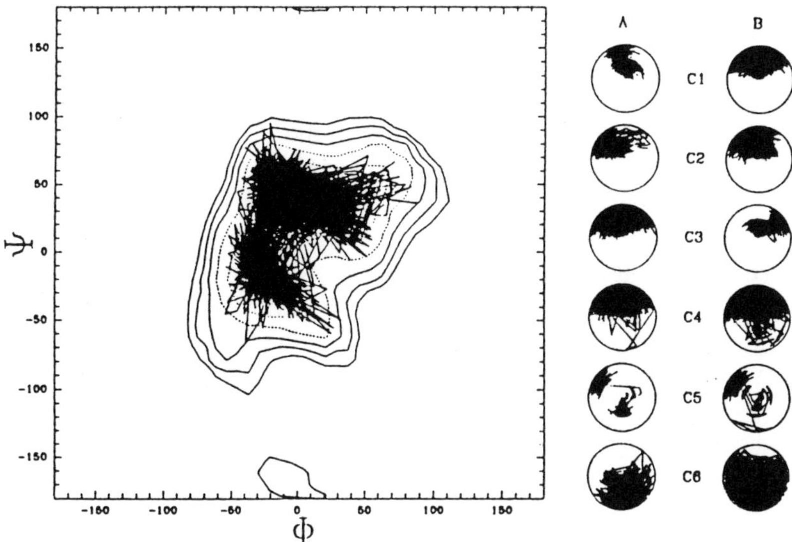

Figure 16. MM2*, GB/SA[26] conformational space of **1** traversed in 5 ns of molecular dynamics from the lowest energy conformation in the C5gt-gg family (+1.70 kJmol⁻¹). φψ space is superimposed upon the gg-gg relaxed map. Conformational dials of torsional trajectories are displayed on the right where A = manα and B = manβ. Conformational dials represent time radially and torsional angle (defined by H-CX-O-H) in a clockface style.[36]

solution properties. A full report on the MM2* GB/SA conformational space of **1** portrayed by molecular mechanic, dynamic and contour map integration is at present in preparation (Figure 16).

V. VERIFICATION OF MOLECULAR MODELING PREDICTIONS

Nuclear magnetic resonance spectroscopy in the form of $^3J_{HH}$ coupling constants and the nuclear Overhauser effect provide a ideal medium to validate molecular modeling studies.

A. $^3J_{HH}$ Coupling Constants: Heparin

Relationships between $^3J_{HH}$ coupling data and molecular conformation such as the Altona-Hasnoot equation are useful in both verifying and aiding modeling studies. The quantity and quality of $^3J_{HH}$ coupling data about pyranose ring systems provides much information about ring geometry. Some of the potential uses of coupling data are well demonstrated by the research history on the iduronic acid ring of heparin which exists in three forms. However the Altona Hasnoot equation was paramatized with coupling data from a set of rigid molecules. This appears to limit its accuracy for the flexible C5–C6 torsions of carbohydrates.[36]

Heparin, a naturally occurring glycosinanioglycan, is the major inhibitor of blood coagulation. It interacts with heparin cofactor (Antithrombin III), accelerating ATIII rate of inhibition of thrombin, Factors IXa, X, XII and Kallikrein (intrinsic coagulation cascade serine proteases) in the order of seven orders of magnitude. This sways the coagulation balance towards fibrinolysis. Heparin is extracted from animal tissues, a process carried out by NZ Pharmaceuticals Ltd. in Palmerston North, New Zealand, and is used clinically as an anticoagulant and antithrombotic.[75]

Although the anticoagulant activity of heparin has been known for seventy years, and ATIII since 1968, their structures and mechanism of interaction are now just being unraveled. The molecular structure of heparin is complex but is largely accounted for by regular sequences of the trisulfated disaccharide **3** [(1-4)-O-(2-O-sulfo-α-L-idopyranosyluronic acid)-(1-4)-O-(2-deoxy-2-sulfamido-6-O-sulfo-α-D-glucopyranose)]. Interspersed on the chain are iduronic acid and glucosamine units containing additional or missing sulfate groups and acetylated groups. Heparin also contains interspersed irregular sequences, one of which is unique, the pentasaccharide **4** which is responsible for the bonding of ATIII[76–79] **4** [O-(2-acetamido-2-deoxy-6-O-sulfo-α-D-glucopyranosyl)-(1-4)-O(β-D-glucopyranosyluronic acid)-(1-4)-O-(2-deoxy-2-sulfamino-3,6-di-O-sulfo-α-D-glucopyranosyl(1-4)-O-(2-O-sulfo-α-L-idopyranosyluronic acid)-(1-4)-O-(2-deoxy-2-sulfoamido-6-O-sulfo-α-D-glucopyranose)]. The sulfate groups are known to be essential for binding to ATIII especially the unique 3-O-sulfo group of residue four.

The interaction between Heparin and ATIII is at present the most understood interaction between a protein and a carbohydrate. Conformational knowledge of ATIII relies upon homologous modeling studies of crystal structures of antitrypsin cleaved at the active site and ovalbulium. The conformational space and modeling of the pentasaccharide sequence of heparin is highly complex as it involves:

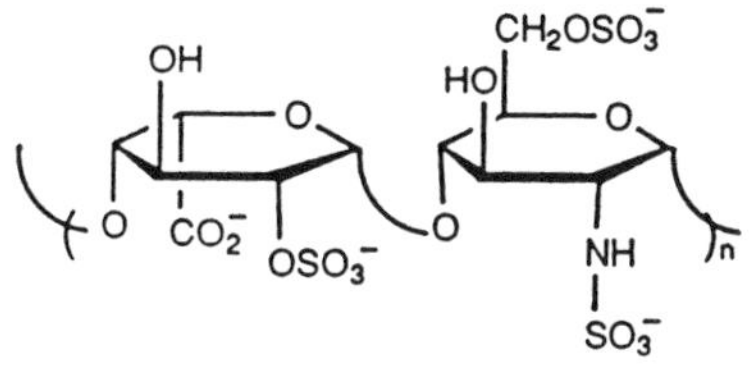

3

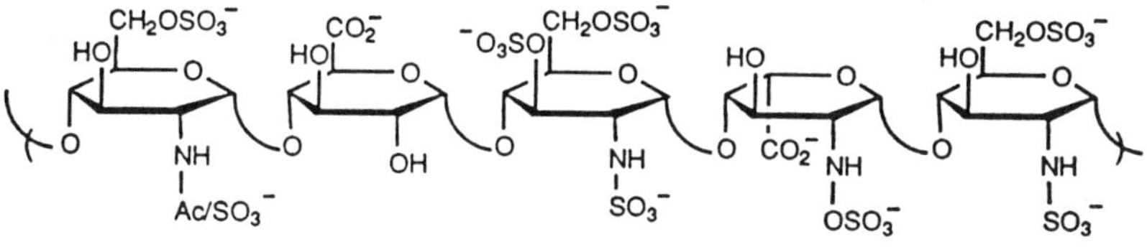

4

- the conformation of the iduronic acid ring which is in equilibrium between three forms,
- the presence of $O\text{-}SO_3^-$ and $NHSO_3^-$ groups for which there are no well established stereochemical data and interaction potentials,
- strong polar interactions, solvent and counter ion effects.

After initial investigation by X-ray diffraction and ^{1}H NMR experiments a $^1C_4$ conformation was assigned to the iduronic acid residue. As the coupling constants are not consistent with a perfect chair it was proposed this chair was distorted.[81,82] In 1981 methyl-4,6-O-(S)-benzylidene-2-chloro-2-deoxy-α-D-idopyranoside was crystallized to a $^2S_0$ skew boat conformation stabilized by all non-H-substituents lying in sterically and electronically favorable positions. Subsequently an equilibrium between the $^2S_0$ and $^1C_4$ ring forms was proposed, with the variation in coupling constants arising from flexibility of the $^2S_0$ form. By driving the ring conformation of methyl-2-O-sulfate-4-methyl-α-L-idopyranose uronic acid using Cremer-Pople coordinates[9] a relaxed type of map determined the sterically allowed $^1C_4$, $^4C_1$, and $^2S_0$ equi-energetic ring conformations and their stability.[85] Each ring conformation is a unique temporal and spatial conformation as barriers hindering the conversion are an order of magnitude. Each ring form is within a relatively steep well indicating no distorted ring conformations contribute to the coupling data, which can therefore only be accounted for by an equilibrium between all three ring conformations. However the map was of an unrelaxed nature and predicted populations based upon relative energies of these and subsequent modeling studies with REFINE were unable to reproduce the experimental coupling data. For this reason all population predictions have been calculated by least square fitting of the coupling constants of the low energy monosaccharides in each ring conformation to experimental data. No full conformational searches of monosaccharide or disaccharide constituents of the pentasaccharide have been reported.

Sequential effects to the iduronic acid ring population were determined by synthesis of at least 30 different sequences. Populations have been calculated as above fitting to appropriate coupling data. The analysis of $^3J_{HH}$ coupling constants has allowed the $^4C_1$ conformation to be excluded from internal sequences. The $^2S_0$ ring contributes up to 40% in internal sequences, 64–69% when preceded by 3-sulfated 2-deoxy-2-sulfoamino residue and is the major contributor 78% when flanked by two 3-O-sulfated amino sugars.[86]

B. The Nuclear Overhauser Effect

The nuclear Overhauser effect (nOe) reflects the relaxation or macroscopic progression towards an equilibrium distribution of nuclear spin states, after an initial perturbation from that equilibrium state. The energy transfer required to flip between spin states requires the simultaneous transfer of an equal quanta of energy to another degree of freedom of the system. The force required to couple a spin transition to the lattice energy levels has the form of a fluctuating local magnetic field arising from the translations, rotations, and internal motions of the molecule in solution. The motion which is to gain or lose energy from the nuclear spin transition must itself cause a local fluctuating magnetic field at the site of that nuclear spin to mediate the energy transfer, to or from, that motion. For this reason the rate of nOe buildup is dependent upon both the conformational (r) and dynamic (τ_c) nature of the molecule in solution.[87]

Links enabling concurrent use of experimental nOe data and molecular modeling techniques in the conformational determination of biological macromolecules are currently attracting worldwide interest. However this relationship between relaxation and molecular motion in solution is complex. The implications of which must be understood to enable its appropriate and accurate application. The remainder of this chapter introduces methods of modeling with nOe restraints and the shortfalls of this method when applied to flexible molecules such as carbohydrates. An alternative method for validating models of carbohydrate molecular motion by calculation and comparison of theoretical nOe to experimental nOe data is reviewed.

Molecular Modeling with nOe Restraints

At present there are various computational strategies that combine nOe distance constraints and various molecular modeling methodology to generate a group of conformations reproducing experimental nOe data. Molecular modeling techniques combine potential energy restraints (r), and energy minimization (rEM), molecular dynamics (rMD) and simulated annealing (SA) procedures. The rationale being that if you repetitively force a system to satisfy enough nOe distance constraints on minimization or dynamic simulation the resultant conformation will satisfy all the nOe restraints and correspond to the ‹actual› solution conformation. The greater number and more accurate these nOe distance constraints are the more accurate subsequent predictions of molecular structures will be. The reliability of the distance constraints depends upon the techniques of extracting distances from nOe spectra which is by no means a trivial problem. Methodology to calculate ^{1}H–^{1}H distances from 2D nOesy spectra has been developed by Kaptein et al. (IRMA), James et al. (MARDIGRAS) and Hare et al. (Hare Research, Inc.).

An Iterative Relaxation Matrix Approach (IRMA[90,91]) has been interfaced into NMRchitect (INSIGHT[92]). IRMA supplements experimental data with theoretical nOe intensities derived from a structural model to enable back transformation of an nOe matrix onto the relaxation matrix. This relaxation matrix, adapted for spin diffusion and simple internal motion effects[93] is used to derive distance constraints. These distance constraints in conjunction with rEM and rMD are used to update the structural model, and refine the NMR restraints. A Procedure for Matrix Analysis of Relaxation for DIscerning GeometRy of an Aqueous Structure, (MARDIGRAS[94]) produces a set of maximum and minimum distance constraints from 2D nOesy data, for subsequent use in conjunction with rEM, rMD and rSA protocols.

These methods were originally designed and utilized for conformational elucidation of proteins. This methodology has been successfully used to investigate the conformational properties of proteins,[95] where in the absence of large scale internal reorientation, the nOe confidently indicates the proximity of sequentially removed residues. However in flexible molecules where there is large scale internal motion we believe that this assumption can lead to misleading information. And therefore we believe that the application of distance restraint methods to carbohydrate systems[96–100] is inappropriate.

All distance restraint models incorporate multispin effects in the calculation of the nOe, that is relaxation of every ^{1}H with every other ^{1}H is taken into consideration. However the above calculations assume isotropic tumbling and that the saccharide tumbles as a rigid body in solution. In other words it is assumed that a single conformation contributes to the solution characteristics of the molecule. The implications of these assumptions are further discussed below. The presence of internal motion, flexibility, conformational averaging (call it what you will) has effects on the nOe through two parameters: (1) the ^{1}H–^{1}H distance (r), a process dubbed conformational averaging, and (2) the spectral density function characterized by τ_c. Effects are due to anisotropic tumbling and internal motion.

Conformational Averaging

Conformational averaging, arises from, and is not dissimilar to, the multiminima problem. It complicates the interpretation of NMR parameters in flexible molecules. An understanding of this problem is essential for accessing the validity of any relationship between NMR data, conformational properties and molecular modeling.

By way of example, consider a population distributed between two conformations (a) and (b) in a ratio 10:90. If conformational exchange between (a) and (b) occurs more slowly than nOe growth, enhancements in the 'separate' (a) and (b) molecules will have time to approach their steady state values uninfluenced by the exchange. On the other hand, if conformational exchange is faster than nOe growth, many exchange events will take place during the time required for the nOe to grow. This allows the spin populations of (a) and (b) to equilibrate and relaxation will in effect reflect an 'averaged molecule'. In this instance consider the ^{1}H–^{1}H distance in (a) to be much less than the corresponding distance in (b) ($r_a < r_b$). This difference is exaggerated in the values of r^{-6} (r_a^{-6} >>>

r_b^{-6}). As the nOe is a function of r^{-6}, this results in the minor conformational population (a) being the main contributor to the observed nOe. Therefore *NMR parameters of systems that have several degrees of internal freedom represent thermodynamically averaged conformational properties which are virtual (i.e., may have no structural significance).*

Conformational Averaging in Carbohydrates

The question 'does conformational averaging occur in carbohydrates?' is analogous to the question of glycosidic flexibility versus rigidity. Coincidental agreement between the solution nOe's of sucrose,[31,101,102] cellobiose,[103] several mannose disaccharides **1** and **2**, Table 1, and simplistic nOe calculations from appropriate rigid geometries, has lent justification to the belief that carbohydrates tumble as rigid bodies in solution. However as this long held belief of carbohydrate rigidity evaporates with more detailed modeling and nOe studies, we propose that methodology assuming rigid body tumbling in solution (IRMA, MARDI-GRAS, FELIX) is inappropriate to the modeling of carbohydrates.

One of the effects of conformational averaging is well demonstrated by nOe modeling studies of the pentasaccharide of heparin have been derived by modeling with nOe distance constraints in a manner similar to those described above. The iduronic acid ring conformation was fixed by least-squares fitting to coupling data as described previously. Distance information was derived from 2D nOesy data under the assumption of a single preferred conformation. An adapted form for MM2, the REFINE force field was utilized. Two low-energy conformations of the pentasaccharide resulted.[98–100] Back calculation of nOe enhancements from the resultant two low energy conformations, each with different iduronic ring conformations, violated several nOe enhancements across the G–A* and I_S–A_M linkages, the latter linkage predicted to be flexible by local relaxed map analysis. This study therefore only serves to reveal that the calculated conformations are virtual in nature and the problem is of much larger dimension than that assumed in this model.

To demonstrate the possible magnitude of error introduced into the conformational analysis of carbohydrates by the assumption of rigid body tumbling, consider the following geometrical analysis. Most nOe data across glycosidic linkages relate linkage carbon ^{1}H's. This corresponds to the F1–G2 ^{1}H–^{1}H distance in **2**. The graphical superimposition

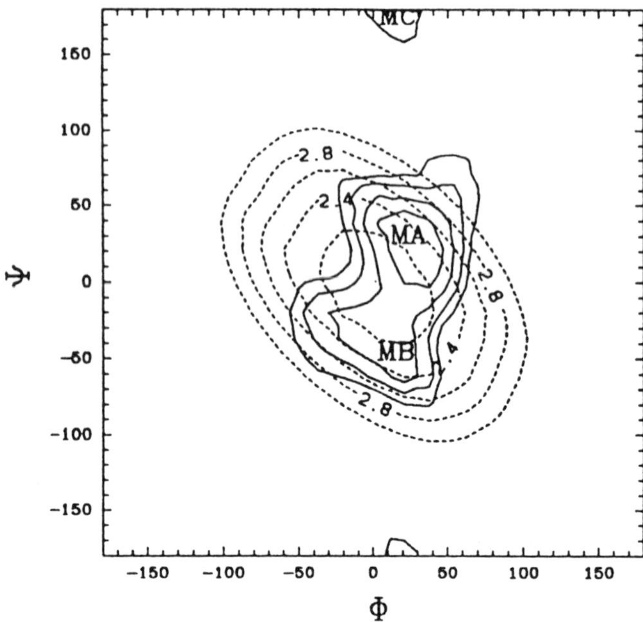

Figure 17. Graphical superimposition of the F1-G2 ^{1}H–^{1}H distance (dashed) superimposed upon the MM2*, GB/SA^{26} relaxed low energy region (solid) of **2**. Contours are 0.2 A and 1 kcal mol^{-1} apart respectively.^{36}

of the F1–G2 ^{1}H–^{1}H distance (dashed) superimposed upon the MM2* GB/SA relaxed low energy region (solid) of **2** is shown in Figure 17. The F1–G2 ^{1}H–^{1}H distance as related to $\phi\psi$ for MA and MB conformations orientate the F1 and G2 ^{1}H's on the same face of the molecule with a ^{1}H–^{1}H distance <3 Å. MC conformations' orientate the F1 and G2 ^{1}H's anti and have ^{1}H–^{1}H distances >3 Å. In this situation the contribution of a MC conformation to the nOe F1–G2 or G2–F1 is at least a factor of 16 times less than the contribution of a MB or MC conformation. Under the assumption that carbohydrates tumble as rigid bodies in solution F1–G2 nOe data is consistent with a conformation in the MA region.^{61} This interpretation however, in the presence of conformational flexibility, could lead to the overlooking a MC population of up to ≈80–90%, which would not be reflected in the nOe.

The consequences of conformational averaging transform the interpretation of nOe data into a formidable problem:^{104}

- From an averaged nOe alone one cannot tell what conformations contribute to this average or even whether there is more than one conformation present. *The nature of these conformations can only come from another source such as molecular modeling.*
- The ability to fit nOe data using predefined conformations *cannot be taken to mean that these conformations are necessarily those present.* This questions the significance of any structure evolved by distance constraint modeling which reproduces nOe contacts.
- It becomes impossible to separate any of the individual conformations contributing to an equilibrium mix without first knowing all possible accessible conformations including the average and associated NMR parameters.

In other words, *in molecules that are undergoing conformational averaging fast on the NMR timescale nOe data can only be used to validate prediction of the dynamic nature of that molecule.*

Validating Molecular Modeling Predictions by nOe

Current methodology, to date, for the comparison of experimental and theoretical nOe data calculated from carbohydrate modeling studies is inaccurate in both modeling and NMR theory. Relative nOe calculations, addressing conformational averaging, have been calculated from relaxed maps (static), and unconverged dynamic systems. Theoretical approaches to relative nOe calculations have assumed isotropic tumbling, r^{-6} averaging and neglected the effect of internal motion on $^{1}H–^{1}H$ correlation time. Previous predictions of theoretical relative nOe (DYNAMO) errors of 20%[67] are substantially underestimated and the effectiveness of the nOe to evaluate modeling predictions of carbohydrates has been questioned[22] and concurrently the legitimacy of any conclusions drawn from comparisons of calculated macroscopic properties from molecular modeling simulations to experimental NMR values must be questioned. In addition reasonable agreement with experimental data for a variety of static and motional models including and excluding conformational averaging effects for the Manα(1-3)Man linkage (Table 1), has led to the questioning of the effectiveness of the nOe to evaluate modeling predictions of carbohydrates.[22]

We are in agreement with Homans that static nOe spectra does not reveal sufficient conformational information to distinguish between different conformational models. However the rate of nOe buildup reveals additional information dependent upon conformation and molecular motion. As the NMR experiment is not subject to any assumptions or restrictions,[105] we believe qualitative simulations of relaxation spectra will succeed when the underlying theory is complete. The rate of nOe buildup is dependent upon both the conformational nature of the molecule in solution and the rate at which it moves between these minima. *Therefore we propose that accurate predictions of nOe buildup rates from molecular motions including treatments of anisotropic tumbling and internal motion will allow classification of differing molecular models.*

To calculate the nOe to a high level of accuracy the dynamic nature of the molecule in solution can be represented as a Fourier transform of the appropriate time correlation function. When internal motion occurs on a substantially faster timescale than molecular tumbling molecular motion can be represented by two independent correlation functions, the first describing the tumbling of a rigid body in solution $(C_{rb}(t))$ and the second the effect of internal motion $(C_{im}(t))$ such that $C(t) = C_{rb}(t)\, C_{im}(t)$. The relative contributions of these correlation times are determined by amplitude factors which depend upon the orientation of each individual ^1H–^1H vector to the main symmetry axis of the cylinder (β), including the effect of internal molecular motions of ^1H–^1H vectors on τ_c. Evaluation of these equations for a cylinder of dimensions of a disaccharide demonstrate the dependence of τ_c on the orientation (β) and internal motion (θ) of ^1H–^1H vectors as shown in Figure 18(a). By way of example an asymmetric molecule, modeled by a cylinder, with the z axis orientated along the longest dimension of the system, Figure 18(a), will tumble at a rate about z ($\theta_{\parallel}$), which is slower than the rate of tumbling about any axis perpendicular to it ($\theta_{\perp}$). This results in the correlation time (τ_c) of a ^1H–^1H vector orientated along the z axis being longer than the correlation time of a ^1H–^1H vector orientated at an angle β (β 0°,180°) to the z axis, Figure 18(b). These variations in τ_c are reflected in nOe buildup rates.

nOe buildup rates for **2** were calculated[36] from snapshots of MD trajectories. The coordinate frame, cylinder dimensions, and tumbling rates were calculated from the average geometry over all MD frames. Most ^1H–^1H vectors of **2** undergo minor internal motion ($\langle\theta_{RMS}\rangle < 0.20$

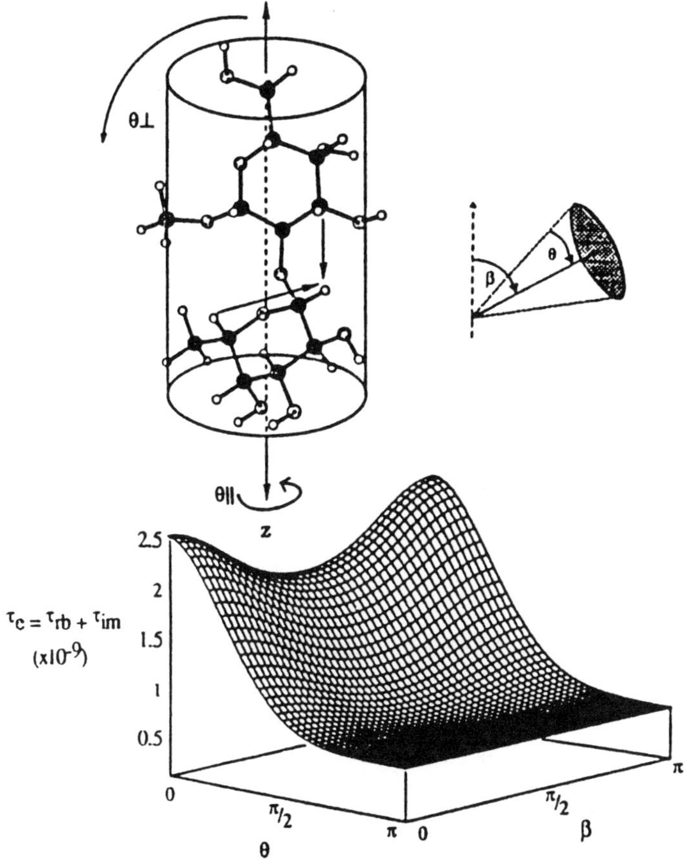

Figure 18. Displays an asymmetric disaccharide **2**, modelled by a cylinder (top). The long axis of this cylinder is defined as the z axis. The cylinder tumbles at a rates $\theta_{\parallel}$ parallel and $\theta_{\perp}$ perpendicular to this long axis. Basic hydrodynamic theory dictates the rate of tumbling to the relevant dimension, therefore parallel tumbling occurs at a faster rate than perpendicular tumbling, $(\theta_{\parallel} > \theta_{\perp})$. The orientation ($\beta$) and internal motion ($\theta$) of ^{1}H–^{1}H vectors therefore effects their effective correlation time (τ_c) (bottom) and nOe.[36]

rads). Internal motion of this degree has little effect on buildup rates and resultant nOe intensities. However, in vectors undergoing fast rotation, such as intra-methyl group ^{1}H–^{1}H vectors, calculations excluding internal motion were in gross error. Treatment of methyl group rotation is complicated by its complex internal motion, and is managed as the superimposition of a bistable jump between three states about a shifting

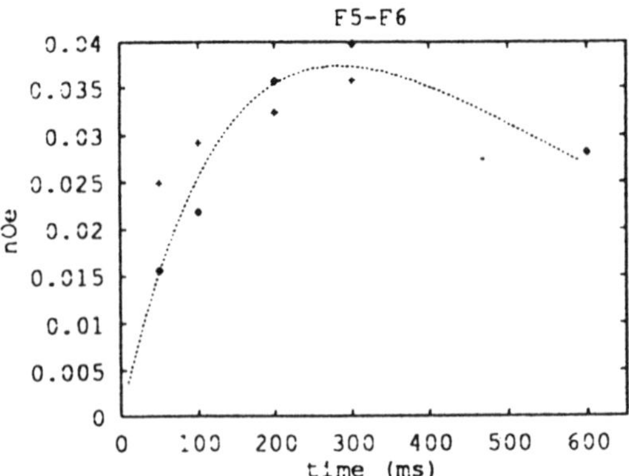

Figure 19. F5–F6 nOe buildup curve of **2**. The dotted line are theoretical values calculated from a molecular dynamic simulation,[107] the points represent 2D nOesy experimental data, ω = 300 MHz, T = 297 K, τ_m = {50, 100, 200, 300, 600}.[36]

axis of rotation relative to the long axis of the cylinder. Assessments of the internal rate of rotation of the F6Me ($\approx 1.4 \times 10^{10}$ s^{-1}) and G1OMe (7.0×10^{10} s^{-1}) methyl groups made from 20 ns MD MM2* GB/SA simulations resulted in good agreement with both the F5–F6 and G1–OMe enhancements (Figure 19).

Anisotropic tumbling results in nOe buildup curves of the slower tumbling ^{1}H–^{1}H vectors orientated parallel to the long axis being of steeper incline than those of ^{1}H–^{1}H vectors orientated perpendicular to the long axis. However the degree of computational complexity introduced by anisotropic tumbling is not without its rewards. The introduction of anisotropic tumbling introduces dependence of the nOe on ^{1}H–^{1}H vector orientation. This therefore provides additional conformational information as different conformational models about the glycosidic linkage orientate ^{1}H–^{1}H vectors at different attitudes to the main axis of symmetry. This is especially useful in carbohydrates as they contain minimal resonances across the glycosidic linkage. For **2** the problem is therefore increased from one in three variables, ϕ, ψ, and the F1–G2 nOe, to one in many variables, ϕ, ψ, G1–G3, G1–G5, F5–F4, F5–F3 and F1–G2 nOe's. The provision of this number and quality of conformationally dependent nOe restraints has allowed the accurate differentiation and evaluation of the different dynamic and mechanic molecular force field and solvent models.

The agreement between calculated and experimental buildup data of the disaccharide D-α-L-fucopyranosyl-(1-2)-O-β-D-galactopyrano-syl-β-O-Me **(2)** using not only the MM2* force field and GB/SA solvent treatment,[106] provides an insight into the rigidity–flexibility controversy about glycosidic linkages, reinforcing the concept of flexibility and supports a complex biological encoding mechanism which may include long range interresidue hydrogen bonding effects.

Methodology to compute the relationship between molecular motion and the buildup of nOe signals encapsulating the effects of conformational averaging on r and τ_c, and the effect of anisotropic tumbling on τ_c, can distinguish between dynamic predictions of conformational space. This therefore provides a new tool for the evaluation of molecular modeling predictions of solution behavior. The development of this methodology and its application,[107,108] provides a link between molecular modeling and NMR spectroscopy and thereby opens a new window of investigation into the conformational dynamic nature of molecules in solution. The method therefore offers potential for a detailed understanding of the behavior of DNA, carbohydrates, and proteins in solution.

REFERENCES AND NOTES

1. Bush, C.A.; Cagas, P. In *Advances in Biophysical Chemistry*, Vol. 2; Bush, C.A. (ed.), JAI Press, Greenwich, CT, 1992, 149–180.
2. Lemieux, R.U. *Pure and Applied Chemistry* **1971**, *25*, 527–548.
3. Brady, J.W. In *Advances in Biophysical Chemistry*, Vol. 1; Bush, C.A. (ed.), JAI Press, Greenwich, CT, 1990, 155–202.
4. Sharon, N.; Lis, H. *Scientific American* **1993**, January, 74–81.
5. It has been asserted by Dewk that, in many cases, the protein portion of a glycoprotein molecule serves as an unimportant and basically nonfunctioning scaffolding whose main purpose is to provide the carbohydrate functional moiety. Dewk, R.A., as quoted in Knight, P. *Biotechnology* **1989**, *7*, 35–40.
6. Okamura, W.H.; Palenzuela, J.A.; Plumet, J.; Midland, M.M. *Journal of Cellular-Biochemistry* **1992**, *49*, 10–18.
7. Jorgenson, W.L. *Science* **1991**, *254*, 954–955.
8. French, A.D.; Brady, J.W. In *Computer Modelling of Carbohydrate Molecules*, French, D.; Brady, J.W. (eds.), ACS Symposium Series: 1990, 1–19.
9. Cremer, D.; Pople, J.A., *J. Am. Chem. Soc.* **1975**, *97*, 1354–1358.
10. Schleifer, L.; Senderowitz, H.; Aped, P.; Tartakovsky, E.; Fuchs, B. *Carbohydr. Res.* **1990**, *206*, 21–39.

11. Venkatachalam, C.M.; Ramchandran, G.N. In *Conformation of Biopolymers*, Vol. 1, Ramachandran, G.N. (ed.), Academic Press, New York, 1971.

12. Lemieux, R.U.; Bock, K.; Delbaere, L.T.J.; Koto, S.; Roo, V.S. *Can. J. Chem.* **1980**, *58*, 631–653.

13. Thogersen, H.; Lemieux, R.U.; Bock, K.; Meyer, B. *Can. J. Chem.* **1982**, *60*, 44–57.

14. Melberg, S; Rasmussen, K. *Carbohyd. Res.* **1980**, *78*, 215.

15. Brady, J.W. *J. Am. Chem. Soc.* **1986**, *108*, 8153–8160.

16. Brooks, B.R.; Bruccoleri, R.E.; States, J.D.; Swaminathan, S.; Karplus, M. *J. Comput. Chem.* **1983**, *4*, 187–217.

17. Nilsson, L.; Karplus, M. *J. Comput. Chem.* **1984**, *1*, 591.

18. Ha, S.N.; Giammona, A.; Field, M.; Brady, J.W. *Carbohdyr. Res.* **1988**, *180*, 207–221.

19. Weiner, P.K.; Profeta, S.; Wipff, G.; Havel, T.; Kuntz, I.D.; Langridge, R.; Kollman, P.A. *Tetrahedron* **1983**, *39*, 1113.

20. Weiner, S.J.; Kollman, P.A.; Case, D.A.; Singh, U.C.; Ghio, C.; Alagona, G.; Profeta, S.; Weiner, P. *J. Am. Ehcm. Soc.* **1984**, *106*, 765–784.

21. Weiner, S.J.; Kollman, P.A.; Nguyen, D.T.; Case, D.A. *J. Comput. Chem.* **1986**, *7*, 230–252.

22. Homans, S.W. *Biochemistry* **1990**, *29*, 9110–9118.

23. Warin, V.; Baert, F.; Fouret, R.; Strecker, G.; Spik, G.; Fournet, B.; Montreuil, J. *Carbohydr. Res.* **1979**, *76*, 11–22.

24. McDonald, D.Q.; Still, W.C. *Tet. Lets.* **1992**, *50*, 7743–7746.

25. McDonald, D.Q.; Still, W.C. *Tet. Lets.* **1992**, *50*, 7747–7750.

26. MacroModel/BatchMin and documentation, version, 3.5X, **1992**.

27. Jeffrey, G.A.; Tayor, R. *J. Comput. Chem.* **1980**, *1*, 99–109.

28. Tvaroska, I.; Perez, S. *Carbohydr. Res.* **1986**, *149*, 389–410.

29. Carver, J.P.; Mandel, D.; Michnick, S.W.; Imberty, A.; Brady, J.W. In *Computer Modelling of Carbohydrate Molecules*, French, D.; Brady, J.W. (eds.), ACS Symposium Series, 1990, 266–280.

30. Tran, V.; Buleon, A.; Imberty, A.; Perez, S. *Biopolymers*, **1989**, *28*, 679–690.

31. French, A.D. *Carbohydr. Res.* **1989**, *188*, 206–211.

32. Ha, S.N.; Madsen, L.J.; Brady, J.W. *Biopolymers* **1988**, *27*, 1927–1952.

33. Brant, D.A.; Christ, M.D. In *Computer Modelling of Carbohydrate Molecules*, French, D.; Brady, J.W. (eds.), ACS Symposium Series, 1990, 42–68.

34. Tran, V.H.; Brady, J.W. *Biopolymers* **1990**, *29*, 961–976.

35. Herve du Penhoat, C.; Imberty, A.; Roques, N.; Michon, V.; Mentech, J.; Descotes, G.; Perez, S. *J. Am. Chem. Soc.* **1991**, *113*, 3720–3727.

36. Coxon, E.E. Unpublished results.

37. Tran, V.H.; Brady, J.W. In *Computer Modelling of Carbohydrate Molecules*, French, D.; Brady, J.W. (eds.), ACS Symposium Series, 1990, 213–226.

38. French, A.D.; Tran, V.H.; Perez, S. In *Computer Modelling of Carbohydrate Molecules*, French, D.; Brady, J.W. (eds.), ACS Symposium Series, 1990, 191–212.

39. Gelin, B.R.; Karplus, M. *J. Am. Chem. Soc.* **1975**, *97*, 6996–7006.

40. Imberty, A.; Tran, V.; Perez, S. *J. Comput. Chem.* **1989**, *11*, 205–216.

41. Ramachandram, G.N.; Sasisekharan, V. *Adv. Prot. Chem.* **1968**, *23*, 283.

42. Tvaroska, I.; Kozar, T.; Hricovini, M. In *Computer Modelling of Carbohydrate Molecules*, French, D.; Brady, J.W. (eds.), ACS Symposium Series, 1990, 162–176.

43. Wiberg, K.B.; Boyd, R.H. *J. Am. Chem. Soc.* **1972**, *94*, 8426–8430.

44. Golden Software, Golden, Colorado.

45. Tvaroska, I. *Int. J. of Quantum Chem.* **1989**, *XXXV*, 141–151.

46. Ragazzi, M.; Provasoli, A.; Ferro, D.R. In *Computer Modelling of Carbohydrate Molecules*, French, D.; Brady, J.W. (eds.), ACS Symposium Series, 1990, 332–344.

47. Tvaroska, I.; Kozar, T.; Hricovini, M. In *Computer Modelling of Carbohydrate Molecules*, French, D.; Brady, J.W. (eds.), ACS Symposium Series, 1990, 162–176.

48. Hooft, R.W.W.; Kanters, J.A.; Kroon, J. *J. Comput. Chem.* **1991**, *12*, 943–947.

49. Starting geometries have been derived prominently from crystal structural data and energy minimized structures from crude conformational search procedures. At the time of writing we know of no fully converged conformation search of a disaccharide reported in the literature.

50. Conformational ANalysis package. Coxon, E.E., **1992**.

51. In addition, the importance of biological motion is overlooked by many X-ray crystallographers and in many modeling studies in attempts to relate structure to function.

52. Resolution of the blurred images of fast moving hooves depicted all four legs in contact with the ground simultaneously.

53. Karplus, M.; Petsko, A. *Nature* **1990**, *347*, 631–639.

54. **2** has recently been synthesized and the crystal structure determined by Dr. David Larsen at the Chemistry Department, University of Otago, New Zealand. The crystal structure is in good agreement with the second lowest energy (+1.57 kJmol^{-1}) MM2* vaccuo conformation, RMS = 0.376 A.

55. Bush, C.A.; Yan, Z.Y.; Rao, B.N.N. *J. Am. Chem. Soc.* **1986**, *108*, 6168–6173.

56. Yan, Z.; Bush, C.A. *Biopolymers* **1990**, *29*, 799–811.

57. Mukhopadhyay, C.; Bush, C.A. *Biopolymers* **1991**, *31*, 1737–1746.

58. Cagus, P.; Bush, C.A. *Biopolymers* **1990**, *30*, 1123–1138.

59. Cagus, P.; Bush, C.A. *Biopolymers* **1992**, *32*, 277–292.

60. Cagus, P.; Kaluarachchi, K.; Bush, C.A. *J. Am. Chem. Soc.* **1991**, *113*, 6815–6822.

61. Kline, P.C.; Serianni, A.S.; Huang, S.G.; Hayes, M.; Barker, R. *Can. J. Chem.* **1990**, *68*, 2171–2182.

62. Homans, S.W.; Dwek, R.A.; Rademacher, T.W. *Biochemistry* **1987**, *26*, 6571–6578.

63. Brisson, J.R.; Carver, J.P. *Biochemistry* **1983**, *22*, 1362–1368.

64. Brisson, J.R.; Carver, J.P. *Biochemistry* **1983**, *22*, 3671–3680.

65. Brisson, J.R.; Carver, J.P. *Biochemistry* **1983**, *22*, 3680–3686.

66. Cumming, D.A.; Carver, J.P. *Biochemistry* **1987**, *26*, 6676–6683.

67. Cumming, D.A.; Carver, J.P. *Biochemistry* **1987**, *26*, 6664–6676.

68. Cumming, D.A.; Dime, D.S.; Grey, A.A.; Krepinsky, J.J.; Carver, J.P. *The Journal of Biological Chemistry* **1986**, *261*, 3208–3213.

69. Cumming, D.A.; Shah, R.N.; Krepinsky, J.J.; Grey, A.A.; Carver, J.P. *Biochemistry* **1987**, *26*, 6655–6663.

70. Winnik, F.M.; Brisson, J.; Carver, J.P.; Krepinsky, J.J. *Carbohydr. Res.* **1982**, *103*, 15–28.

71. Dime, D.D.; Rachaman, E.; Dime, C.E.; Grey, A.A.; Carver, J.P.; Krepinsky, J.J. *J. Labelled Compounds and Radiopharm.* **1987**, *XXIV*, 725–739.

72. Homans, S.W.; Pastore, A.; Dwek, R.A.; Rademacher, T.W., *Biochemistry* **1987**, *26*, 6649–6655.

73. Noggle, J.H.; Schirmer, R.E. *The Nuclear Overhauser Effect.* Academic Press, New York, **1971**.

74. Edge, C.J.; Singh, U.C.; Bazzo, R.; Taylor, G.L.; Dwek, R.A.; Rademacher, T.W. *Biochemistry* **1990**, *29*, 1971–1974.

75. Jacques, L.B., *Pharmacol. Rev.* **1980**, *31*, 99.

76. Choay, J.; Lormeau, J.; Petitou, M.; Sinay, P.; Fareed, J. *J. Ann. NY Acad. Sci.* **1991**, *370*, 664.

77. Thunberg, L.; Backstrom, G.; Lindahl, U. *Carbohydr. Res.*, **1982**, *100*, 393.

78. Choay, J.; Petitou, M.; Lormeau, J.; Sinnay, P.; Casu, B.; Gatti, G., **1983**, *116*, 492.

79. Sinay, P.; Jacquinet, J.; Petitou, M.; Duchaussoy, P.; Lederman, I.; Choay, J.; Torri, G. *Carbohydr. Res.* **1984**, *132*.

80. Gatti, G.; Casu, B.; Perlin, A.S. *Biochemical and Biophysical Research Communication* **1978**, *85*, 14–20.

81. Gatti, G.; Casu, B.; Torri, G.; Vercellotti, J.R. *Carbohydr. Res.* **1979**, *68*, c3–c7.

82. Gatti, G.; Casu, B.; Hamer, G.K.; Perlin, A.S. *Macromolecules* **1979**, *12*, 1001–1007.

83. Chamberlain, L.N.; Edwarads, I.A.S.; Stadler, H.P.; Buchanan, J.G. *Carbohydr. Res.* **1981**, *90*, 131–137.

84. Casu, B; Choay, J.; Ferro, D.R.; Gattis, G.; Jacquinei, J.P.; Provasoti, A.; Ragazzi, M.; Sinay, P.; Torri, G. *Nature* **1986**, *322*, 215–216.

85. Ragazzi, M.; Ferro, D.R.; Provasoli, A. *J. Comput. Chem.* **1986**, *7*, 105–112.

86. Ferro, D.R.; Provasoli, A.; Ragazzi, M.; Torri, G.; Casu, B.; Gatti, G.; Jacquinet, J.C.; Sinay, P.; Petitou, M.; Choay, J. *J. Am. Chem. Soc.* **1986**, *108*, 6773–6778.

87.

$$f_1\{S\} = \frac{\gamma_S}{\gamma_I} \frac{W_{2IS} - W_{0IS}}{W_{0IS} + W_{1I} + W_{2IS}} \qquad W_{0IS} = Kr_{IS}^{-6}\left[\frac{\tau_c}{1 + (\omega_I - \omega_S)^2\tau_c^2}\right]$$

$$W_{1I} = \frac{3}{2} Kr_{IS}^{-6}\left[\frac{\tau_c}{1 + \omega_I + \omega_S^2\tau_c^2}\right] \qquad W_{2IS} = 6Kr_{IS}^{-6}\left[\frac{\tau_c}{1 + (\omega_I + \omega_S)^2\tau_c^2}\right]$$

where $K = \gamma^4 h^2 = 5.7 \times 10^4$ s, ω_X is the Larmour frequency of spin X, r_{IS} is the average $^1H_I - {}^1H_S$ distance in nm, and τ_c is the effective correlation time. For a homonuclear system the transition probabilities can be simplified as $\gamma_I = \gamma_S$, $\omega_I = \omega_S = \omega$, $|\omega_I = \omega_s| = 0$ and $\omega_I + \omega_S = 2\omega$.

88. Won, H.; Olson, K.D.; Hare, D.R.; Wolfe, R.S.; Kratky, C.; Summers, M.F. *J. Am. Chem. Soc.* **1992**, *114*, 6880–6892.

89. Summers, M.; South, T.L.; Hare, D.R. *Biochemistry* **1990**, *29*, 329–340.

90. Boelens, R.; Koning, T.M.G.; Kaptein, R.J. *J. Molec. Struct.* **1988**, *173*, 299.

91. Boelens, R.; Koning, T.M.G.; van der Marel, G.A.; Van Boom, H.; Kaptein, R.J. *J. Magn. Res.* **1989**, *82*, 290.

92. Insight II and documentation, Biosym Technology, Inc., San Diego, CA, USA.

93. Koning, T.M.G.; Boelens, R.; Kaptein, R. *J. Magn. Reson.* **1990**, *90*, 111–123.

94. Borgias, B.A.; James, T.L. *J. Magn. Reson.* **1990**, *87*, 475–487.

95. Thomas, P.D.; Basus, V.J.; James, T.L. *Proc. Nat. Acad. Sci. USA* **1991**, *88*, 1237–1241.

96. Krishna, N.R.; Choe, B.Y.; Harvey, S.C. In *Computer Modelling of Carbohydrate Molecules*, French, D.; Brady, J.W. (eds.), ACS Symposium Series, 1990, 227–239.

97. Choe, B.Y.; Ekborg, G.C.; Roden, L.; Harvey, S.C.; Krishna, N.R. *J. Am. Chem. Soc.* **1991**, *113*, 3743–3749.

98. Ragazzi, M.; Ferro, D.R.; Perly, B.; Sinay, P.; Petitou, M.; Choay, J. *Carbohydr. Res.* **1990**, *195*, 169–185.

99. Ferro, D.R.; Provasoli, A.; Ragazzi, M.; Casu, B.; Torri, G.; Bossennec, V.; Choay, J. *Carbohydr. Res.* **1990**, *195*, 157–167.

100. Ragazzi, M.; Ferro, D.R.; Perly, B.; Torri, G.; Casu, B.; Sinay, P.; Petitou, M.; Choay, J. *Carbohydr. Res.* **1987**, *195*, c1–c5.

101. Herve du Penhoat, C.; Imberty, A.; Roques, N.; Michon, V.; Mentech, J.; Descotes, G.; Perez, S. *J. Am. Chem. Soc.* **1991**, *113*, 3720–3727.

102. Tran, V.H.; Brady, J.W. *Biopolymers* **1990**, *29*, 961–976.

103. French, A.D.; Tran, V.H.; Perez, S. In *Computer Modelling of Carbohydrate Molecules*, French, D.; Brady, J.W. (eds.), ACS Symposium Series, 1990, 191–212.

104. Neuhaus, D.; Williamson, M. *The Nuclear Overhauser Effect in Structural and Conformational Analysis*, VCH Publishers, Inc., USA, 1989.

105. Duben, A.J.; Hutton, W.C. *J. Am. Chem. Soc.* **1990**, *112*, 5917–5924.

106. MacroModel v3.5. Mohamadi, F.; Richards, N.G.J.; Guida, W.C.; Liskamp, R.; Lipton, M.; Caufield, C.; Chang, G.; Hendrickson, T.; Still, W.C. *J. Comput. Chem.* **1990**, *11*, 440–467.

107. CAMMALFMFCR the Unpronounceable. Program for the CorrelAtion of Molecular Motion Antecedent to the Local Fluctuating Magnetic Fields Causing Relaxation. Coxon, E.E., **1992**.

108. Withka, J.M.; Swaminathan, S.; Beveridge, D.L.; Bolton, P.H. *Science* **1992**, 255, 597–599.

THE MOLECULAR MODELING POTENTIAL ENERGY SURFACE

D. Ross Boswell, Edward E. Coxon,

and James M. Coxon

Advances in Molecular Modeling
Volume 3, pages 195–224.
Copyright © 1995 by JAI Press Inc.
All rights of reproduction in any form reserved.
ISBN: 1-55938-326-7

I. INTRODUCTION

The past two and a half decades have seen the evolution and rapid development of computational chemistry. The term computational chemistry encompasses molecular modeling, semiempirical, and ab initio methods. The term molecular modeling refers to techniques utilizing a force field empirical approximation to the potential energy surface, that is, molecular mechanics and molecular dynamics. Scientific theory has as its ultimate goal the understanding and emulation of natural phenomena. Embracing this goal computational chemistry constructs a model of the real world from which both measurable and immeasurable properties are computed. Comparison of the measurable properties to experimentally determined properties enables an assessment of the validity of the model used, and its immeasurable predictions.

Due to the complex and unhomogeneous nature of chemical systems, approximations are required to reduce the number of degrees of freedom to where computer simulation becomes possible. An important part in the art of computer simulation is to choose the approximations of the molecular model and computational procedure such that their contributions to the overall accuracy are of comparable size, without affecting significantly the property of interest. The three major problems in the molecular modeling of complex systems are (1) the potential energy function problem, (2) the solvation problem, and (3) the multiminima problem.

This chapter focuses on the unavoidable assumptions, approximations and simplifications involved, their effect on the accuracy of the predictions and defines the assumptions implicit to molecular mechanic and dynamic integration.

II. THE POTENTIAL ENERGY PROBLEM

A. Mathematical Description

As atoms within a molecule move with respect to each other the stability or relative energy of that molecule alters. This energy surface can be compared to the countryside. Molecules exist in the valleys or low energy, low stress, regions. As the number of atoms (N) within the molecule increases so too does the dimensionality ($3N - 6$) of the countryside, which begins to contain tunnels, bridges, hyper-space by-ways, and more convoluted topographical features. The mathematical description of this potential surface is a formidable problem concerning both small particles and large velocities, hence quantum mechanical and relativistic effects become important. Erwin Schrödinger earlier this century developed the time independent wave equation which has become known as the Schrödinger Wave Equation (SWE) based on atomic structure theory proposed by Bohr, and wave mechanics of de Broglie.

$$HY(\mathbf{R},\mathbf{r}) = EY(\mathbf{R},\mathbf{r}) \tag{1}$$

The SWE is a quantum mechanical description of the interaction of the set of particles in a molecule, the electrons, neutrons and protons. It relates the "wave function," which describes the spatial distribution of the particles as a function of the coordinates of the electrons ($\mathbf{r}$) and nuclear particles ($\mathbf{R}$), to the linear operators for kinetic and potential energy (E). The SWE is too complex for practical application as the integral can only be solved for small systems ($< {}^4$He) of no interest to experimental chemists and biochemists.

The *Born–Oppenheimer (BO) approximation* forms the theoretical basis for computer modeling of molecular potential energy surfaces. As the energy in atoms is equi-partitioned between constituent parts, the electrons which are several thousand of times lighter than the nuclei move much faster, allowing essentially instantaneous electronic compensation for underlying nuclei movements. Thus the motion of the electrons can be decoupled from the motion of the nucleus, enabling separation of the SWE into two independent equations. Equation (2) defines a potential energy surface (E) set up by the electronic fields that adapt almost instantaneously to any change in the nuclear coordinates. Equation (3) describes the motion of the nuclei on this potential energy surface, represented by the nuclear wave function, according to forces set up by the electronic fields represented by the potential energy surface.

$$\mathbf{H}\Psi(\mathbf{r};\mathbf{R}) = \mathbf{E}\Psi(\mathbf{r};\mathbf{R}) \tag{2}$$

$$\mathbf{H}\Phi(\mathbf{R}) = \mathbf{E}\Phi(\mathbf{R}) \tag{3}$$

Quantum mechanical calculations freeze the position of the nuclei, place the electrons in the lowest unfilled molecular orbitals (basis set) and obtain the atomic wave function as an iterative eigenvalue problem. The matrix diagionalization procedure required to solve this equation is time consuming, resulting in a time dependence of approximately n^4 for ab initio methods, where n is the number of atomic orbitals. Therefore, quantum mechanical calculations are limited to small molecules ($n <$ 20). Semiempirical calculations, discussed in chapter 2 of this volume, are n^3 time dependant as only valance electrons are included explicitly, with empirical parameters being used for core orbital contributions.[2]

B. Molecular Mechanics

Molecular mechanics invokes radical empirical approximations to model the detailed forces within molecules created by their electronic structure, Eqs. (2), (3). This results in a computationally efficient description of the potential energy surface with a time dependence of approximately N^2 where N is the number of atoms. However as the size and flexibility of molecules increase computational requirements increase at a rate $>>N^2$ as the multiminima problem places further efficiency demands on energy evaluation.[3]

Recent computer and algorithm development has enabled all atom representations of the dynamic behavior (<100 ps) of systems in solution containing up to $N < 2000$ atoms. For example, a 100 ps simulation of a DNA dodecamer in one thousand and twenty seven water molecules took 40 hours CPU on a CRAY supercomputer.[4] Molecular modeling is therefore currently the method of choice for macromolecular structures such as proteins, DNA, oligosaccharides, and even for medium sized molecules.

In the molecular mechanics approach outlined by Westheimer in 1956,[5] a molecule is considered as a collection of particles held together by simple elastic or harmonic forces. These forces can be described by a set of empirical penalty potential energy functions of the internal coordinates of the molecule, such as bond lengths, bond angles, and dihedral angles. A form of the Westheimer equation is,

$$E_{\text{total}} = E_{\text{stretch}} + E_{\text{bend}} + E_{\text{van der Waals}} + E_{\text{nonbonded}} + \cdots$$

where E_{total} is the "steric energy." *The potential energy is the difference between the real molecule and a hypothetical molecule where all struc-*

tural features are at their ideal values. As the energy is only a measure of intramolecular strain relative to a hypothetical situation, it does not have any physical meaning. Therefore, comparisons of energies are only valid on the same potential energy surface, that is, differing conformations of the same molecule.

The superimposition of these force field equations forms a surface in $3N-6$ dimensions, where N is the number of atoms. The potential surface of test molecules is parameterized or contorted, twisted and scaled until the observed minimum for each internal coordinate overlaps the minimum of experimentally observed geometries. Thus it is hoped that close to a minima energy structure the force field approximates the potential energy surface, and the potential energy surface of known molecules can be extrapolated to the unknown.

The MM2* and AMBER* Potential Forms

Since Westheimer's proposal of the potential form, many different force fields have evolved. All partition the potential energy surface into different groups of structural features and can use varying numerical methods to calculate the energy contribution from each of these terms. Early trends in force field development were towards basic force field equations for simplicity, understanding and computational speed. However it is becoming apparent that more complicated potential forms are required in order to obtain self consistent fields. By way of example, original difficulties fitting vibrational frequencies to MM2(77) and conformational inaccuracies can be overcome by fitting several additional terms to the force field equation set. However the required revision of MM2 would be so extensive that the new force field MM3 has been born. However, at present the parameterization for MM3 is in its infancy. Consequently, for the next few years MM2 will remain the most viable tool for force field analysis of organic molecules.[6] For proteins the AMBER force field is the most viable tool. There are many modeling packages, for example MacroModel,[7] BAKMOD,[8] and INSIGHT,[9] all of which implement various force fields. The more prominent force fields are Allinger's family of force fields [MM1,[10] MM2(77),[11] MM2(87),[12] and MM3[13]]; modified versions of Kollman's AMBER,[14–16] GROMOS[17] and CHARMM.[18,19] MacroModel force fields are denoted as MM2* and AMBER*.[20,21]

An understanding of each individual force field term in the MM2 and AMBER force fields as with any force field gives an insight into the nature of the potential energy surface. This is necessary since it allows an appropriate choice of force field to be made for a particular class of compounds. A brief discussion of a form of the MM2 and AMBER potentials follows.

A MM2* Potential Form

$$E_{MM2} = E_s + E_b + E_{sb} + E_t + E_{nb} + E_c + E_{hb} \qquad (4)$$

$$E_s = \sum_{i=1}^{n} k_i^s (l_i - l_{i0})(1 - 2.0(l_i - l_{i0})) \qquad (a)$$

$$E_b = \sum_{i<j}^{m} k_{ij}^b (\theta_{ij} - \theta_{ij0})^2 \left(1 + 7.0 \times 10^{-8}(\theta_{ij} - \theta_{ij0})^4\right) \qquad (b)$$

$$E_{sb} = \sum_{i<j} k_{ij}^{sb}(\theta_{ij} - \theta_{ij0})(l_i - l_{i0} + l_j - l_{j0}) \qquad (c)$$

$$E_t = \sum_{T} (\tfrac{1}{2} V_i)(1 + \cos \phi) + \tfrac{1}{2} V_2(1 - \cos 2\phi) + \tfrac{1}{2} V_3(1 + \cos 3\phi) \qquad (d)$$

$$E_{nb} = \sum_{i<j} \varepsilon * \left(-c_1 \left(\frac{r_{ij}^*}{r_{ij}}\right)^6 + c_2 \exp\left(-c_3 \frac{r_{ij}^*}{r_{ij}}\right)\right) \qquad (e)$$

$$E_c = \sum_{i<j}^{m} \frac{q_i q_j}{cr_{ij}} \qquad (f)$$

$$E_{hb} = \sum_{HB} \left(\frac{C_{ij}^*}{r_{ij}^{12}} - \frac{D_{ij}^*}{r_{ij}^{10}}\right) \qquad (g)$$

An AMBER* Potential Form

$$E_{AMBER} = E_s + E_b + E_t + E_{nb} + E_c + E_{hb} \qquad (5)$$

$$E_s = \sum_{i=1}^{n} k_i^s (l_i - l_{i0})^2 \qquad \text{(a)}$$

$$E_b = \sum_{i<j}^{m} k_{ij}^b (\theta_{ij} - \theta_{ij0})^2 \qquad \text{(b)}$$

$$E_{sb} = \sum_{i<j} k_{ij}^{sb} (\theta_{ij} - \theta_{ij0})(l_i - l_{i0} + l_j - l_{j0}) \qquad \text{(c)}$$

$$E_t = \sum_{T} (\tfrac{1}{2}V_1)(1 + \cos\phi) + \tfrac{1}{2}V_2(1 - \cos 2\phi) + \tfrac{1}{2}V_3(1 + \cos 3\phi) \qquad \text{(d)}$$

$$E_{nb} = \sum_{HB} \left(\frac{C_{ij}^*}{r_{ij}^{12}} - \frac{D_{ij}^*}{r_{ij}^{10}} + \frac{q_i q_j}{\varepsilon r_{ij}} \right) \qquad \text{(e)}$$

$$E_{hb} = \sum_{HB} \left(\frac{C_{ij}^*}{r_{ij}^{12}} - \frac{D_{ij}^*}{r_{ij}^{10}} \right) \qquad \text{(f)}$$

The *bend* (E_b [Eqs. (4), (5a)]) and *stretch* (E_s [Eqs. (4), (5b)]) steric energy contributions to the system are summed over all bonds (n) and bond angles (m) of the system, where l_{i0} and q_{ij0} are the ideal bond lengths and angles for atoms i and j, l_i and q_{ij} are the actual bond lengths and bond angles, and k^s and k^b are the stretching and bending constants.

The energy required to stretch or bend a bond from its ideal value is approximated by Hook's law (Figure 1). This, however, overestimates the energy required to achieve large distortions from the ideal values. For example as an alkane C–C–C bond angle is distorted far from its natural value of 108°23′ ie. towards 90° or 180° the hybrid sp^3 atomic orbitals do not overlap well. Therefore less energy is required for large deformations than Hook's law would predict. This illustrates the need for more complex terms to approximate energy contributions from structural features. A cubic term [Eq. (4a)] was added to the Hook's law quadratic term of MM1 in MM2. This allows a more accurate assessment of the energy of strained molecules where large deviations occur. However this can also result in the "blowing apart" of bad starting geometries

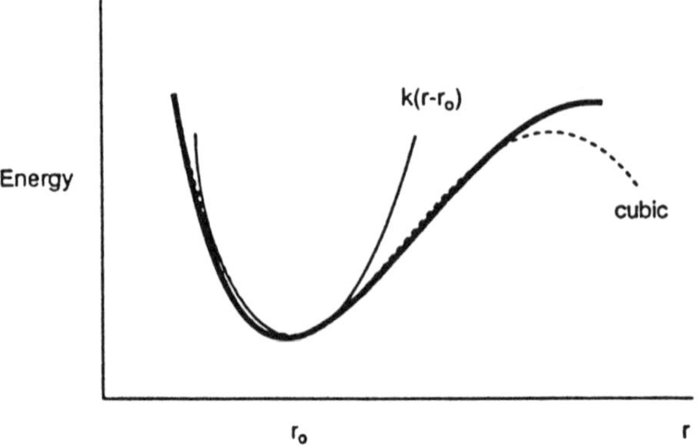

Figure 1. The energy required (bold) to stretch or bend a bond from its ideal value (r_o) is approximated by Hook's law with and without the inclusion of a cubic term.

during minimization, and during dynamics, when bond angles or lengths move far from equilibrium onto the cubic part of the function. This problem can be avoided during dynamics by switching off the cubic term, and reduced during minimization by turning off the cubic term if initial bond lengths and angles are far from equilibrium values. This does not however completely avoid the problem during torsional angle fixing. In this situation the cubic term is not deactivated as the initial bond lengths and angles are within the ideal range, however on minimization bad steric interactions can force large deviations of bond angles and lengths onto the cubic function, resulting in forces which dissipate the structure.[22] The *stretch bend* term (E_{sb} [Eq. (4c)]) was introduced into MM2 to model the bond length increase that occurs when the corresponding bond angle decreases. The energy barrier for *torsional rotation*[23] (E_t [Eqs. (4), (5d)]) is described by a truncated Fourier series to three terms. V_1, V_2, and V_3 are parameterised force constants.

In general it takes many times less energy to distort a dihedral angle from its ideal value (for single bonds) than it does to bend a bond. In turn it takes a factor of ten times more energy to stretch a bond than to bend it. Hence when strain has been averaged over a molecule, distortions tend to be present in torsional angles > bond angles > bond lengths.

The pairwise nonbonded interaction potential (E_{nb} [Eq. (4e)]) is summed over all the unique pairs of atoms that are not bonded to each other or to a common third atom. Any 1,3 interactions are incorporated into the bend term. Both torsional and nonbonded treatment are required for 1,4 interactions. The analytical form of the nonbonded potential E_{nb} is comprised of two forces, attractive London dispersion electrostatic forces at long range and van der Waals repulsion at close range. MM2 uses the Hill equation to approximate London dispersion forces (r^{-6} dependant) and an exponential function for van der Waals repulsion, where ε^* the "hardness" of each atom is related to the gradient of the repulsive segment of the potential well calculated from $(\varepsilon_k\varepsilon_i)^{-0.5}$ and ε_k is an atomic parameter for atom k, r_{ij}^* is the sum of the van der Waals radii of the interacting atoms ($r_i + r_j$), r_{ij} is the effective interatomic radii, and c_{1-3} are universal constants.

The uneven distribution of electron density between partially charged atoms is approximated by a Coulomb term (MM2*), [E_c Eqs. (4f), (5e)] or by dipole–dipole electrostatics (MM2). All force fields contain default partial charge parameters. More accurate partial charges can be obtained from ab initio or semiempirical calculations for the molecule under consideration. The selection of atomic charges in simulations of solutions of hydrogen bonded solutes in dipolar solvents, such as water is difficult. The dynamic changing specific polarization of each hydroxyl group resulting from alterations in its instantaneous environment, is in general represented as an average by a single fixed partial atomic charge.[24]

The computation requirements of nonbonded interactions is proportional to N^2 where N is the number of atoms. As all other potential energy terms are time dependant on N, computational efficiency can be dramatically improved by reduction of the nonbonded interaction list. There are many different treatments of long range electrostatic interactions to reduce the nonbonded interaction array size, which are well reviewed by van Gunsteren et al.[25] For example, constant distance dielectric electrostatics with infinite cut-offs and a nonbonded interaction array update time of 0.1 ps can be used. Such a treatment enhances calculation rates by the definition of a nonbonded interaction array of neighboring atoms lying within a cut-off which are updated periodically, and avoids noise originating from the discontinuity of nonbonded interactions around a cut-off radius which artificially increases the kinetic energy of the atoms.

Hydrogen bonding is considered to be largely an electrostatic phenomenon. The hydrogen bond is short (<5 Å), of a high energy (<20 kJ mol^{-1}), and is angularly dependant. Hydrogen bonding is often of considerable importance. For example, it is a significant conformational force in carbohydrates as they contain many freely rotating hydroxyl groups. The energy of hydrogen bond formation is large relative to the value of the 'Boltzmann constant $\times$ room temperature' and therefore can have a significant effect on relatively flat potential energy surfaces. For molecules capable of forming hydrogen bonds computational methods that neglect hydrogen bonding can result in inaccurate predictions of conformational space. For example the formation of a hydrogen bond transforms an otherwise relatively high energy conformation of sucrose into a highly populated minimum.[26]

MM2(77) contained no explicit hydrogen bond treatment and most of the energy contribution due to hydrogen bonding was accounted for by the electrostatics. However ab initio calculations revealed somewhat shorter and stronger ($1–3$ kcal mol^{-1}) hydrogen bonds than calculated by MM2(77). Subsequently MM2(87)[12] incorporated a hydrogen bonding term in the form of a van der Waals type energy function [Eq. (4e)], as it has the correct functional characteristics, accounts for the angular dependence of hydrogen bonding, and was simple to implement. The van der Waals radius of the hydrogen atom involved in the hydrogen bond is reduced for that interaction only. MM2* uses Kollman's angle independent 6,12-Lennard–Jones treatment for hydrogen bonding. (E_{hb} [Eqs. (4g), (5f)]) where each HB pair (H·X) are assigned Lennard–Jones C and D parameters and are summed over all interactions in the force field.

Parameterization of a Force Field

Parameterization of a force field occurs in two phases. First, the ideal values and steric energy penalty for deviations from these ideals are defined. Second, the entire parameter set is 'massaged' in order to obtain a self consistent force field that correctly reproduces experimentally observed conformational preferences.[27] This massage compensates for experimental parameterization data error, scales all interactions to a common zero point and incorporates the quantum nature of the system into the analytical form.

It is easy to transcend the boundary between legitimate and illegitimate use of parameters. For this reason the quality and origin of all parameters utilized in a system must be investigated and reported. As strain deviates torsional angles to a greater extent than bond angles and lengths, resultant structures and their relative energies are most sensitive to torsion parameters.

III. BOUNDARY CONDITIONS, THE SOLVATION PROBLEM

The most interesting phase for chemists, biochemists and biologists is the solution phase. Modeling approaches in this phase face two additional problems, namely

- the inclusion of solvent effects into calculations significantly increases the number of degrees of freedom of the system, and
- evaluation of the modeling methodology is difficult as in the solution phase few accurately measurable atomic properties are available.

Water molecules form a hydration shell around charged polar molecules on the molecular surface, altering charge distribution and offering hydrogen bond donors and acceptors. The water bound to the molecule has very different properties from water in the bulk state. Bound water has a much slower mobility, longer resonance times, and slower exchange rates than nonbound water.

The effect of solvation on dynamic fluctuations is a question of considerable debate. Dielectric screening of charges would be expected to *increase flexibility*. Adjacent water molecules are spatially restricted, and structured around certain groups of the solute in such a way that each water molecule can still make hydrogen bonds to other water molecules, and avoid the high energetic cost of loosing a hydrogen bond.[28] The structuring of the solvent shell around the molecule will *decrease* large torsional fluctuations. However, atoms can gain kinetic energy by random collision with these solvent molecules, enhancing conformational exchange.

Solvent effects on molecules which contain many groups capable of forming hydrogen bonds are expected to be pronounced as the electrostatically least stable structures (in molecules containing polar functional groups) are typically the most heavily solvated in a polar solvent. In polyhydroxy compounds the dependence of conformation on intramolecular hydrogen bonding will be less pronounced in solution than in vacuo calculations or in the crystal structure.

There are many modeling techniques designed to address the solvation problem. They are classified as vacuo, continuum and explicit boundary conditions. These classes of treatment, thoroughly reviewed elsewhere,[25] are briefly introduced below.

Calculations in vacuo correspond to gas phase boundary conditions at zero pressure. The shielding effect of water on electrostatic interactions between charges and/or dipoles is lacking in vacuo. Therefore, calculations in vacuo tend to distort molecules as the surface area is minimized.

Continuum boundary conditions analytically approximate the effect of solvation.[29,30] The analytical form has easily differentiated derivatives for use in conjunction with the molecular modeling force field. This statistical continuum approach provides small molecule hydration energies of comparable accuracy to those obtained from free energy perturbation methods[31] at a fraction of the computational expense, enabling solvation calculations that take approximately twice as long as similar calculations in vacuo.[30] Solvation energies will be accurate only if high quality atomic partial charges are used. The best method to compute partial charges for a new system is to carry out high level quantum mechanical calculations and fit atomic partial charges to reproduce the electric field at the solvent accessible surface.

Periodic boundary conditions can be used to explicitly include solvent molecules within identical cubic translated space filled boxes, that is, 3^3-1^3 boxes for the first shell, $5^3-3^3-1^3$ boxes for the second shell, and so forth. Explicit inclusion in this manner is efficient but still results in, CPU intensive calculations (CPU $\propto N^2$ therefore simulations are 100–1000 times longer than in vacuo calculations), simulation length and size limitations, and convergence problems in both molecular mechanic minimizations and molecular dynamic simulations.

IV. INTEGRATION OF THE POTENTIAL ENERGY SURFACE: THE MULTI–MINIMUM PROBLEM

The multi-minimum problem is considered the most difficult aspect of molecular modeling and arises from:

- the complexity of conformational hyperspace.
- the nature of molecules in solution at nonzero temperatures.
- the methods for exploring this space—molecular mechanic and molecular dynamic integration.

A. The Conformational Hyperspace

Early computer modeling methods reduced the degrees of freedom of systems by predicting molecular properties at 0 K, ie. global minima or crystal conformations (Figure 2, MB).

However, at *nonzero temperatures* kinetic energy moves a molecule over its conformational space. This is represented in Figure 2 by movement from MB through barriers A, C and E to MD and MF. For chemical systems in solution at nonzero temperatures the many molecules are distributed over their conformational space. The different methods of describing this distribution have important consequences which are discussed in this chapter, and further discussed in the previous chapter on molecular modeling of carbohydrates.

As the temperature of a system increases so too does the conformational space accessible to the molecule. Or in other words, as temperature increases the solution properties of a system are less well represented by the global minima. This is most pronounced in flexible systems. By way of example, consider a protein system which undergoes reversible thermal denaturation. As the temperature of the system increases higher

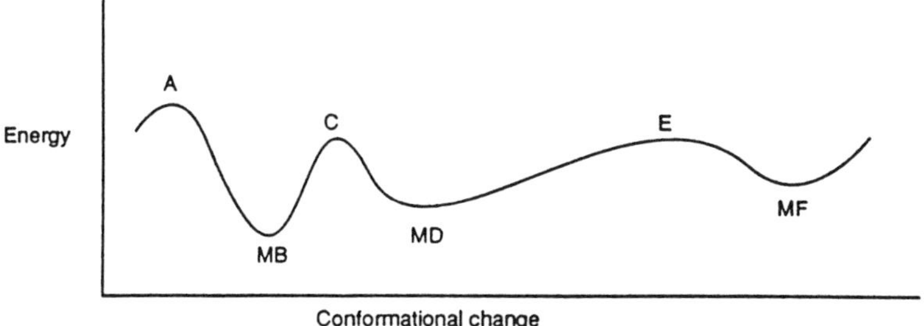

Figure 2. A simplistic diagrammatic representation of the relative energy within a molecule with respect to conformational change.

energy states become accessible. In a protein system with many degrees of freedom there will be a large number of these higher energy states— imagine an unfolded protein! Even though the population in each of these high energy minima is small unfolded conformations become the major contributor to solution properties. For this reason the simulation of nonzero solution properties of a flexible molecule cannot simply be reduced to a few degrees of freedom ie. global minimum. Representation and classification of all low energy regions of the potential surface is required. This is referred to as the multiminima problem.

To convey the complexity of this problem consider the dimensionality of the potential surface requiring exploration. The conformational hyperspace of a molecule, consisting of N atoms, spans $3N - 6$ dimensions, whose axes are the $3N - 6$ degrees of freedom expressed in terms of the internal coordinates (bond lengths, bond angles, torsional angles). A simple disaccharide such as D-α-L-fucopyranosyl-(1-2)-O-β-D-galactopyranosyl-β-O-Me (**1**), has 67 atoms, hence the potential energy surface has $3(N - 6) = 195$ mutually dependant degrees of freedom (dimensions) on which all low energy regions require description and representation.

Before discussion the molecular mechanic and dynamic methods for the location of all low energy regions it is useful to consider the theoretical treatments available for evaluating the contribution of differing low energy regions of the potential surface, or in other words methods for calculating an integral of the potential surface at any given temperature. It is important to recognize the implicit differences between molecular mechanic and molecular dynamic integrals.

1

B. Statistical Mechanics

The Boltzmann order principle provides a link between the macroscopic and microscopic behavior of systems by introducing probability not as a approximation but as an explanatory principle.[32] This enables identification of patterned behavior within macroscopic systems by virtue of them being composed of a large population to which the laws of probability can be applied. The number of ways of distributing N particles into N_1 and N_2 partitions is defined as the number of complexions (P), where

$$P = \frac{N!}{N_1! N_2!} \tag{6}$$

This enables the conceptualization of the macroscopic behavior of systems composed of many particles. For any given population the smaller the difference between N_1 and N_2 the larger the number of complexions. The number of complexions is maximum when the population is distributed evenly over the two halves. Moreover, the larger the value of N the greater the difference between the number of complexions corresponding to the different ways of distribution. Hence for the order of N found in macroscopic systems (i.e., $N = 10^{23}$) the overwhelming majority of possible distributions corresponds to the distribution $N_1 = N_2 = N/2$. This signifies that for systems composed of a large number of particles, all states that differ from the state corresponding to that of equal distribution are highly improbable.

In a closed system defined by boundary conditions such that the temperature (T) is kept constant by heat exchange with the environment, Boltzmann's equilibrium is not defined by maximum entropy but in terms of the minimum of free energy (G).

$$G = H - T\delta S \tag{7}$$

Equilibrium then becomes a balance between two factors, the *enthalpy* (*H*) and the *entropy* (*S*). The balance being defined by the temperature (*T* K). At low temperature enthalpy prevails and the formation of *ordered* (weak entropy) structures is favored, that is, crystal structures and global minima conformations, MB (see Figure 2). At high temperatures entropy or molecular disorder is dominant. The importance of relative motion is increased as systems progress from solid to liquid to gas and due to the large number of accessible states MD will become the highest populated minima. The entropy of an isolated system and the free energy of a system at fixed temperature are examples of thermodynamic potentials. The extremes of these thermodynamic potentials H and S define the attractor states towards which systems whose boundary conditions correspond to the definitions of these potential's will tend simultaneously. We live in a 'luke warm' world where bioorganic chemistry occurs in the solution phase, where properties of systems are determined by a balance between disorder (S) and order (H), with order predominating [Eq. (7)].

C. Integrals of the Potential Surface

Predictions of macroscopic behavior of molecular systems at nonzero temperatures can be made from a statistically representative set of conformations, *a so-called ensemble average or integral over accessible conformational space.* This section describes the molecular mechanic and molecular dynamic generation of these integrals.

Molecular Mechanics

Molecular mechanics (conformational search/minimization) approximates an integral over conformational space by a Boltzmann distribution [Eqs. (8), (9)] of all low energy conformations located, weighted as a function of the enthalpy (H) at 0 K rather than the free energy at room temperature (300 K).

$$P_i = \exp\left[\frac{-H_i}{kT}\frac{1}{Q}\right] \tag{8}$$

where P_i is the relative population of state i ($P_i = 1$) and Q is the classical partition function given by

$$Q = \sum_i \exp\left[\frac{-H_i}{kT}\right] \tag{9}$$

Thus, the underlying assumptions of invariant enthalpy change with temperature and the entropic equivalence of different conformations are implicate to the molecular mechanical approach. These assumptions are to some extent justifiable in small inflexible molecules as the enthalpy only undergoes a small linear variation (related to the heat capacity) with temperature change and the entropic contribution to the free energy is similar for all conformations. As molecular size and flexibility increase neglect of entropic degrees of freedom of the molecule are likely to become a source of increasing error.

The entropic contribution to the free energy of a minima energy conformer can be approximated by *Rigid Rotor Harmonic Oscillator normal vibrational mode analysis*. This however assumes that the potential energy surface (i.e., shape of the well) is of an harmonic nature, which is seldom appropriate especially in larger systems. This renders the RRHO method highly unreliable.

In addition, molecular mechanics ignores the relationship between low energy minima on the potential energy hyper-surface. In certain circumstances this can lead to a poor understanding of the conformation properties of the molecule.[21,34] Therefore to depict the conformational behavior of a molecule at nonzero temperatures it is not only desirable to determine the low energy conformations but also the shape of these wells, their inter-relationships and the energy of transition states between these low energy conformations. Saunders has proposed a small kick stochastic method of searching for conformational inter-conversions to determine the stability and inter-relationships of low energy conformations located upon energy surfaces.[35] This problem however, is of greater dimensionality than the location of all low energy regions on the potential energy surface as a larger proportion of conformational space requires examination.

Molecular Dynamics

Entropy is a difficult quantity to evaluate within the molecular modeling framework. Boltzmann equated entropy (S) to the number of ways (complexions P) of arranging a molecule into a given state [Eq. (10)] where k_b is Boltzmann's proportionality constant.

$$S = k_b \log P \qquad (10)$$

In other words the entropic (S) contribution to energy for a conformation can be related to the number of accessible states within that well, which in turn is related to the shape of each well. By way of example consider a system where the global minimum is in a relatively steep well MB compared to the next lowest minimum which is a broad well of slightly higher energy MD (Figure 2). The broader well has more accessible conformational states and therefore is entropically stabilized becoming the most densely populated minima, with temperature.

Molecular dynamics predicts the movement of a molecule over its potential surface at a given temperature. In other words the potential energy surface is used as a potential guide for molecular motion (termed Potential of Mean Force, PMF). In the flatter regions of the potential surface (i.e., MD), there is less force upon atoms and as a result they travel "slower" and therefore are represented to a greater extent. For this reason conformational subdivisions that can be arranged in a relatively large number of possible ways are represented to a greater extent. Thus *fully converged molecular dynamic simulations implicitly include entropic effects.*

An integral of conformational space can be generated by taking 'geometrical snapshots' of the simulation trajectory at regular intervals. The structures collected or resultant integral will represent the region of conformational space over which the molecular dynamic trajectory traversed. This introduces an important point of issue, the time frame of conformational space coverage.

The rate of exchange between any two low energy regions of conformational space is related to the relative height of the corresponding transition states. Thus the length of real time simulation required to traverse conformational space is dictated by the level and number of transition barriers in the molecule. The simulation time possible is a function of the size of the system, the computer power available and the time step. The time step is inversely proportional to the computational time required for any given length of real time simulation. The balance between decreasing the time step for increased accuracy and increasing the time step for greater simulation duration is a multi-factorial decision.[9,30] To summarize, the time required to cover conformational space is dependant upon the size of the system and the activation barriers to rotation present. In addition, to obtain statistically significant integrals of the traversed conformational space each transition must have occurred a number of times.

It has recently been accepted that molecular dynamics is less efficient than conformational search routines at generating initial starting geometries for subsequent minimization to all low energy conformations on any given potential energy surface.[39,40] That is, although molecular dynamics is the equivalent of an internal coordinate search which is biased to the low energy regions of conformational space, it traverses this space less efficiently than conformational search procedures moving between minima at a slower rate. The status of computer and computer algorithm design has in the past limited molecular dynamic solvent simulation lengths of medium sized molecules such as disaccharides to real times of <500 ps[36] and those of larger molecules such as a DNA dodecamer to 100 ps.[4] This time frame generally limits molecular dynamic simulations to local conformational space coverage. However the introduction of continuum solvent models makes all atom molecular dynamic simulations on disaccharides of 20 ns accessible on modest computers such as IBM RISC 6000 320H.[37]

By fully depicting the conformational space of several disaccharides by both molecular mechanic and molecular dynamic simulations we have been able to implicate errors in molecular mechanical statistical integration methodology[37] (see also chapter on molecular modeling of carbohydrates in this volume). Demonstrating that conformational search algorithms depict conformational space at least an order of magnitude faster than molecular dynamic simulations does not automatically imply that conformation searching is the method of choice for determining the solution properties of small to medium sized flexible molecules.

V. ALGORITHMS FOR THE INTEGRATION OF THE POTENTIAL SURFACE

The remainder of this chapter gives an overview of the algorithms for the generation of both molecular mechanic and dynamic exploration of conformational space.

A. Molecular Mechanic Conformational Search Simulations

Molecular mechanics generates an integral over the potential surface by combining conformation search procedures with minimization algorithms. This locates a set of low energy minima conformations. There is no general solution to the global minimization of a multi dimensional problem as minimization algorithms can only start with a geometry on the potential energy surface and progress "downhill" descending to the nearest local minimum. Consider the universe of all possible conformations of a molecule as an imaginary landscape (Figure 2). If you are in a valley (MB) you cannot see beyond the adjacent hilltop (A and C). Minimization, or walking downhill, from any point between A and C will result in MB. To locate another valley (MD or MF) you have to climb over a hill (A or B) to explore the unknown beyond. Subsequently there is no way of knowing that B is only a local minimum and not the global solution until additional minimizations with starting geometries past C are investigated (i.e., to locate MD and MF). *Therefore conformational searching is an iterative procedure requiring a method of generating a set of conformations that sample the potential energy surface*, to allow subsequent minimization to all the low energy conformational wells.

Internal Coordinate Searches

Distance geometry methods explore the potential energy surface in cartesian coordinates,[38] distance geometry,[15] or internal coordinate[39–41] reference frames. *The overall effectiveness of the search in locating all low energy conformations (i.e., reaching convergence), is determined by the mechanism of generating the set of conformations that sample the conformational space.* Subsequent discussion is limited to *internal coordinate search algorithms* as there is considerable theoretical and experimental[37,39–41] evidence of their greater efficiency exploring conformational space.

The $3N-6$ degrees of freedom of a molecule are spanned by its internal coordinates (bond lengths, bond angles, and torsional angles) of which, due to the parabolic nature of the potential energy function for stretch and bend functions, the bond length and bond angle degrees of freedom are constrained. Thus utilizing internal coordinates the problem of searching the $3N-6$ dimensional conformational space is reduced to one in torsional space, having $n - 1$ dimensions, where n is the total number of torsions. *With torsional bonds being relatively more flexible than bond lengths and bond angles internal coordinate reference systems not only reduce the dimensionality of the problem but reduce it to the most relevant and most unconstrained variables.*

The problem of searching conformational space is still one of significance, by way of example consider **1**. If each torsional angle is restricted to gauche or anti conformations this alone yields $M^n = 3^{18}$ or 3.87×10^8 possible conformers requiring consideration, where M is the number of points sampled along each torsional coordinate, and n is the number of torsions. In this simplistic calculation a considerable proportion of these geometries do not place the ring ends close enough to meet and many have excessive energy due to nonbonded interactions, reducing the total number of conformations considerably. But in turn there are also many conformers which have one or more torsions which depart substantially from ideal *gauche* or *anti* geometries, creating additional conformers for consideration. This has been well illustrated by the failure of deterministic internal coordinate searches with 120° resolution ($M = 3$), to locate all low energy minimum.[41]

Grid searches deterministically cover conformational space by considering all combinations of torsional rotations of every torsion at a given resolution, in other words casting a net over the potential landscapes. The *Internal Coordinate Search Tree*[39] improves the efficiency of the grid search by imposing geometric tests to exclude high energy conformations. Conformational space is divided into a tree like arrangement where each point grows M branches so the number of branches at level i is given by the product $M_1 \cdot M_2 \cdot M_3 \cdot M_4 \cdots M_{i-1}$. If any unfavorable interactions are encountered early on in the search tree a whole branch of that tree can be selectively pruned in one operation. Calculation time is also reduced as geometries formed in previous branches remain fixed. This methodology enables grid searches of molecules with of up to 10^6 and 10^9 possible conformations for acyclic and cyclic molecules, respectively. The *Unbounded Systematic Search* SUMM[41] enables deterministic coverage of conformational space of up to 10^{14} possible conformations for cyclic molecules. The search begins at low torsional angle resolution (120°) and proceeds to high resolution (30°). With each jump in resolution the volume of conformational space being searched expands exponentially so at the highest resolution achieved the search is likely to be incomplete. However, locating favorable conformational space at low resolution enables exploration of the more promising regions at higher resolution first. This method is 30–60% faster than MCMM for small molecules.

Grid searches enable confidence in the location of all low energy conformations. However, grid searches are only applicable to small systems. Deterministic searches of medium to large flexible structures, generate too many starting geometries for current computing capacity even at harsh resolution. For example a conformational search of the backbone of a small protein (50 amino acids 150 torsions) at 120° resolution would keep a supercomputer fully occupied for more than 10^{20} years.

Modeling algorithms based upon evolution principles have lead to the development of search strategies which increase the efficiency of *Monte Carlo internal coordinate search algorithms*[40] (MCMM). The processes central to evolution as proposed by Darwin are natural selection, determining which individuals survive to reproduce, and sexual reproduction, the mixing of genes. Translated into the language of complex multidimensional hyperspace, successful species (relatively low energy structures) are generally more closely related to other successful species (relatively low energy structures) than to unsuccessful species (relatively high energy structures). Monte Carlo search algorithms are in essence genetic algorithms,[42] utilizing these two essential features of evolution to locate the most successful organisms, that is, the lowest relative energy conformations. These algorithms follow the precedent of natural selection, and used the more successful, lower energy structures, as parents. From these parents offspring are generated by Monte Carlo methods, random variation of torsional arrays. This corresponds more to genetic mutation that to conformational sex. Search protocols based on the above observations intelligently bias search protocols toward low energy regions of torsional space. Monte Carlo searches reliably cover the conformational space of flexible structures of up to 12 torsions. To confidently conclude convergence multiple searches from differing starting geometries with varying energetic bounds are required to locate no new low energy minimum.[21,30,40]

Minimization

Molecular mechanics explores the energy surface of a molecule through conformational searches and energy minimization procedures, to find stable points, where the net force on each atom vanishes. This results in a theoretical '0 K' structure.

The minimization procedure is iterative with each iteration involving three phases: (1) the choice of a direction vector, (2) functional evaluation of the energy of the molecule at a point along the direction vector, and (3) adjustment of the internal coordinates of the molecule to the coordinates that correspond to the lower value of the potential function.

The method of direction vector generation determines the minimization procedure, and its efficiency. The efficiency is related to the number of iterations required, and the number of functional evaluations required per iteration, to reach convergence criteria. The minimization algorithm most likely to result in minimization to a low gradient most quickly depends on both the size, harmonics, and initial conditions of the system. A brief description of the various minimization methods for generating direction vectors, and their advantages and disadvantages follows.

Steepest descent minimizations directly equate the gradient to the direction vector at a particular point in conformational space. This reliance upon the gradient results in both the strength and weakness of this method. Far away from a minimum where the molecule is highly strained and the surface is highly anharmonic, steepest descent is the minimization algorithm most likely to result in convergence. Close to a minima, however, the gradient is small and convergence is slow. As each direction vector is perpendicular to the last direction vector the path followed oscillates towards the minimum retracing itself. This inefficient behavior is especially prominent on energy surfaces with narrow valleys. In conjunction with the line search,[43] steepest descent is more robust when surfaces are highly anharmonic, i.e. far from equilibrium.

Conjugate gradient[44] methods prevent retracing earlier minimization progress by generating a complete basis set of mutually conjugate directions and mutually orthogonal gradients, that span the conformational space.[45] The new direction vector $\mathbf{d}_{i+1}$ from the point $i + 1$ is calculated by adding the gradient at point $i + 1$ ($\mathbf{g}_{i+1}$) to the previous direction scaled by γ_i.

$$d_{i+1} = g_{i+1}\gamma_i d_i \tag{11}$$

where γ_i is a scalar defined by

$$\gamma_i = \frac{g_{i+1} \cdot g_{i+1}}{g_i g_i} \tag{12}$$

As this set of conjugate directions span the energy surface then by definition minimization along each must result in the minimum. For a harmonic system it takes $3N - 6$ steps to converge. However, as the derivation invokes a quadratic approximation to the anharmonic nature of the potential energy surface several iterations in each direction may be required for convergence. In certain situations the system can even minimize along conjugate directions without convergence.

The *Full Matrix Newton Raphson* method utilizes the second derivative of the potential energy surface to predict where the gradient will change direction. For harmonic systems the inverse of the second derivative matrix can be multiplied by the gradient to obtain a vector that translates directly to the nearest minimum.

$$\mathbf{r}_{min} = \mathbf{r}_0 - A_0 \cdot E(\mathbf{r}_0) \tag{13}$$

where $\mathbf{r}_{min}$ is the predicted minimum, $\mathbf{r}_0$ is an arbitrary starting point, A_0 is the Hessian matrix of the second partial derivatives of energy with respect to coordinates at $\mathbf{r}_0$, and $E(\mathbf{r}_0)$ is the gradient of the potential energy surface at $\mathbf{r}_0$.

If the gradient is small, the surface is close to being harmonic and the Full Matrix Newton Raphson algorithm is very efficient, requiring only a small number of iterations. However, as the anharmonicity of the potential energy surface increases the method must be applied iteratively. If the gradient is not small, this method is very sensitive due to the inversion of the Hessian matrix. For example, when the forces are large and the curvature small (on the repulsive section of van der Waal's interaction), the algorithm computes a large step (large gradient/small curvature), which may overshoot the minimum and lead to a point even further away. This can cause rapid divergence if the initial forces are too high (or the potential surface is too flat). For these reasons full matrix

Newton Raphson minimizations must be used with line searching if the initial gradient is >0.5 kJ/Å/mol.

Newton Raphson minimizations are slow as it is computationally expensive to calculate the Hessian matrix from the force field equation format. Solving this $N{\times}N$ matrix results in $3N^2$ words of data, resulting in the computational limit $N < 250$, where N is the number of atoms. The full matrix Newton Raphson algorithm is the method of choice for final convergence to a low gradient for small systems already close to minimum.

B. Molecular Dynamic Simulations

Molecular dynamics[46] enables a dynamic picture of molecular behavior over time to be extracted from the force field and parameter set created for molecular mechanical calculations. As previously mentioned the SWE can be divided into two separate equations on application of the BO approximation. Equation [3] describes the motion of the nuclei on the potential energy surface, due to the electronic force field. Solving this equation is of interest in the study of time evolution of a molecule. Molecular dynamics approximates Eq. (2) by the empirical force field and parameter set and Eq. (3) by Newton's equation of classical motion ($F = ma$) which then becomes

$$\frac{d\mathbf{r}_i(t)}{dt} = m_i^{-1}\mathbf{F}_i(\{\mathbf{r}_i(t)\}) \tag{14}$$

where the force $\mathbf{F}$ acting on an atom i in a molecular system is the negative gradient of the empirical force field (E) with respect to atomic coordinates ($\mathbf{r}_i$), and m_i is the mass of atom i.

Classical physics only produces closed solutions to this equation for systems of one or two independent particles. To enable computation for the large systems of interest this integral is approximated numerically using a Taylor series. If the position of atom i ($\mathbf{r}_i(t)$) at time t is known then the position after time interval t is

$$\mathbf{r}_i(t+\Delta t) = \mathbf{r}_i(t) + \frac{d\mathbf{r}_i(t)}{dt}\Delta t + \frac{d^2\mathbf{r}_i(t)}{dt^2}\frac{t^2}{2} + \cdots \tag{15}$$

To complete the dynamic picture of a molecule all that is required is the initial starting coordinates $\mathbf{r}_i(t)$, the velocity ($\delta \mathbf{r}_i/\delta t$), the acceleration ($\delta^2 \mathbf{r}_i/\delta t$) and suitable approximations to account for contributions from higher terms of the Taylor series.

Dynamic simulations are initiated from a starting geometry, minimized in the appropriate force field. Velocities $\mathbf{v}_i(t)$ are assigned to each atom randomly selected from a Boltzmann distribution at the appropriate temperature, and the derivative of the empirical force field is equated to the acceleration on each atom. The system can then be integrated over time by a variety of integration algorithms.

In the derivation of the integration algorithms the acceleration and velocities are assumed to change linearly over any given time step t. Consider this example, when two interacting atoms are far apart the energy change is small, as are the forces, so the atoms move slowly in the set time step therefore portraying near linear velocity and acceleration. As they descend towards the minimum their velocities increase, taking larger and larger steps until they reach the minimum. At this point of high velocity the forces are changing the most therefore atoms step through the energy barrier momentarily. Their speed is rescaled immediately afterwards but not before a small increment is irreversibly added to the energy integral. To justify the assumptions of linear velocity and acceleration over the periods of the time step, each vibrational period must be split into 5–10 time steps. The fastest vibrational frequency is the C–H stretch which is in the order of 10^{-14} Hz, and corresponds to time steps in the order of 10^{-15} sec or 1 fs. The iterative SHAKE algorithm[47] constrains C–H and O– lone pair bond lengths so a larger time step can be used. When simulating macromolecular systems without constraints, a time step of 0.5 fs is appropriate, whereas if hydrogen and lone pair bond lengths are constrained with the SHAKE algorithm a time step of 1 fs will result in accurate energy predictions.[9,30] SHAKE results in a two times increase in real time simulation rate for a 10% increase in computational time.

C. Stochastic Dynamic Simulations

Stochastic dynamics is an extension of molecular dynamics. A trajectory of a molecular system is generated by integration of the stochastic

Langevin's equation of motion which includes a stochastic force $m_i\mathbf{R}_i(t)$ and a frictional force $_i\mathbf{v}_i(t)$.

$$\frac{d\mathbf{r}_i(t)}{dt} = m_i^{-1}\mathbf{F}_i(\{\mathbf{r}_i(t)\}) + m_i^{-1}\mathbf{R}_i(t) + {}_i\mathbf{v}_i(t) \tag{16}$$

where $_i$ is the frictional coefficient of atom i. The condition for zero energy loss is $\langle\mathbf{R}_i^2\rangle = 6m_{ii}k_bT_{\text{ref}}$, where T_{ref} is the reference temperature.

Thus stochastic dynamics mimics solvent effects by establishing a coupling of the individual atomic motions to a heat bath. Keeping atomic kinetic energy nearly constant in this manner greatly increases rates of conformational exchange. The stochastic term represents collisions of solute atoms with solvent molecules. The frictional term represents the drag exerted by the solute on solvent internal motion.

VI. SUMMARY

The choice of potential surface described by the force field and solvation treatment is essential to the molecular modeling of a system. So to is the method of integration of this potential surface for realistic predictions of molecular systems. It should be noted that a "best" force field, solvation treatment, conformational search procedure, minimization algorithm, or dynamic simulation, does not exist. The most appropriate force field and model will depend on the type of molecular system and the type of properties one is interested in.[25,48]

It is therefore important to be aware of the fundamental assumptions, simplifications and approximations that are implicate in the various models used in the literature.

REFERENCES AND NOTES

1. The term computational chemistry encompasses molecular modeling, semiempirical, and ab initio methods. The term molecular modeling refers to techniques utilizing a force field empirical approximation to the potential energy surface, that is, molecular mechanics and molecular dynamics.
2. The less complex the modeling procedure the more experimental calibration is required in the form of paramitization, to reproduce useful accuracy. Ab initio methods with large basis sets require no paramitization. Semiempirical methods

require limited paramitization, while molecular modeling requires full paramitization.

3. By way of example, the derivative of the potential energy surface with respect to atomic coordinates must be calculated for a fully converged conformational search or molecular dynamics simulation in the order of 6×10^6 and 2×10^7, respectively, for a molecule containing approximately 12 flexible torsions.

4. Swaminathan, S.; Ravishanker, G.; Beveridge, D.L. *J. Am. Chem. Soc.* **1991**, *113*, 5027–5040.

5. See: Engler, E.M.; Andose, J.D.; Schleyer, P.v.R. *J. Am. Chem. Soc.* **1973**, *95*, 8005–8024.

6. Lipkowitz, K.B. *QCPE Bulletin* **1992**, *12*, 6–11.

7. Mohamadi, F.; Richards, N.G.J.; Guida, W.C.; Liskamp, R.; Lipton, M.; Caufield, C.; Chang, G.; Hendrickson, T.; Still, W.C. *J. Comput. Chem.* **1990**, *11*, 440–467.

8. BKMDL, MODEL and PCMODEL are molecular modeling packages from Serina Software, CA, USA.

9. Insight II and documentation, Biosym Technology, Inc., San Diego, CA, USA.

10. Allinger, N.L. *Adv. Phys. Org. Chem.* **1976**, *13*, 1.

11. Allinger, N.L. *J. Am. Chem. Soc.* **1977**, *99*, 8127–8134.

12. Allinger, N.L.; Kok, R.A.; Imam, M.R. *J. Comput. Chem.* **1988**, *9*, 591–595.

13. Allinger, N.L.; Yuh, Y.H.; Lii, J.H. *J. Am. Chem. Soc.* **1989**, *111*, 8551–8575.

14. Weiner, P.K.; Profeta, S.; Wipff, G.; Havel, T.; Kuntz, I.D.; Langridge, R.; Kollman, P.A. *Tetrahedron* **1983**, *39*, 1113.

15. Weiner, S.J.; Kollman, P.A.; Case, D.A.; Singh, U.C.; Ghio, C.; Alagona, G.; Profeta, S.; Weiner, P. *J. Am.Chem. Soc.* **1984**, *106*, 765–784.

16. Weiner, S.J.; Kollman, P.A.; Nguyen, D.T.; Case, D.A. *J. Comput. Chem.* **1986**, *7*, 230–252.

17. van Gunsteren, W.F.; Berendsen, H.J.C.; *GROMOS86: Groningen Molecular Simulation System*, University of Groningen, The Netherlands, **1986**.

18. Brooks, B.R.; Bruccoleri, R.E.; States, J.D.; Swaminathan, S.; Karplus, M. *J. Comput. Chem.* **1983**, *4*, 187–217.

19. Nilsson, L.; Karplus, M. *J. Comput. Chem.* **1984**, *1*, 591.

20. McDonald, D.Q.; Still, W.C. *Tet. Lets.* **1992**, *50*, 7743–7746.

21. McDonald, D.Q.; Still, W.C. *Tet. Lets.* **1992**, *50*, 7747–7750.

22. Boyd, D.B.; Lipkowitz, K.B.; *Journal of Chemical Education* **1982**, *59*, 269–274.

23. The torsional angle a-b-c-d is defined as the angle measured about the b-c axis from the abc plane to the bcd plane. Several differing definitions are possible. For example, MacroModel defines clockwise rotation when looking from b-c as positive. Torsion ranges can also vary from –180° to 180°, or from 0° to 360°.

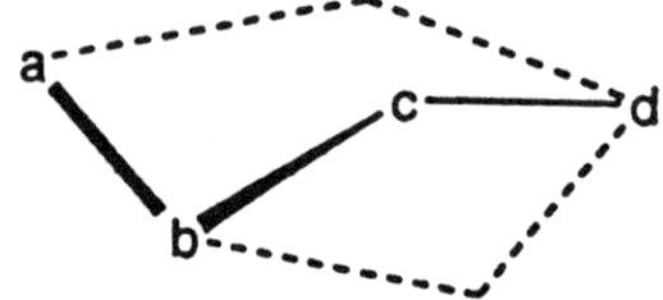

24. Ha, S.N.; Giammona, A.; Field, M.; Brady, J.W. *Carbohydr. Res.* **1988**, *180*, 207–221.

25. van Gunsteren, W.F.; Breendsen, J.C. *Agnew Chem. Int. Engl.* **1990**, *29*, 992–1023.

26. Tran, V.H.; Brady, J.W. In *Computer Modelling of Carbohydrate Molecules*, French, D.; Brady, J.W. (eds.), ACS Symposium Series, 1990, 213–226.

27. Pearlman, D.A.; Kollman, P.A. *J. Am. Chem. Soc.* **1991**, *113*, 7167–7177.

28. Madsen, L.J.; Ha, S.N.; Tran, V.H.; Brady, J.W. In *Computer Modelling of Carbohydrate Molecules*, French, D.; Brady, J.W. (eds.), ACS Symposium Series, 1990, 69–90.

29. Hasel, W. Hendrickson, T.F.; Still, W.C. *Tetrahedron Computer Methodology* **1988**, *1*, 103–116.

30. MacroModel/BatchMin programs and documentation v 3.5X, **1992**.[7]

31. Still, W.C.; Tempczyk, A.; Hawley, R.C.; Hendrickson, T. *J. Am. Chem. Soc.* **1990**, *112*, 6127–6129.

32. The essential mission of Newton's traditional physics has been to reduce the complexity of nature to the simplicity of elementary behavior expressed by the laws of motion. This belief has limited man's conceptualization of the world for the last three hundred years, rendering him unable to identify with many patterns of behavior present in complex systems. Prigogine, I.; Stengers, I. *Order Out of Chaos-Man's New Dialogue with Nature*, Williamn Collins Sons and Co., Glasgow, Scotland, 1984.

33. Newton's causal determinism is confronted as Boltzmann's interpretation implies that a system, whatever its initial conditions, will be attracted to the state of maximum symmetry. Whatever the evolution precursor to the system it will be ultimately be attracted to one of the macroscopic states corresponding to the macroscopic state of disorder and maximum symmetry and probability.

34. Anet, F. *J. Am. Chem. Soc.* **1990**, *112*, 7172–7178.

35. Saunders, M. *J. Am. Chem. Soc.* **1987**, *109*, 3150.

36. Edge, C.J.; Singh, U.C.; Bazzo, R.; Taylor, G.L.; Dwek, R.A.; Rademacher, T.W. *Biochemistry* **1990**, *29*, 1971–1974.

37. Coxon, E.E. *B. Med. Sci. Thesis*, University of Otago, New Zealand, 1993.

38. Saunders, M.J. *J. Am. Chem. Soc.* **1987**, *109*, 3150.

39. Lipton, M.; Still, W.C. *J. Comput. Chem.* **1988**, *9*, 343–355.

40. Chang, G.; Guida, W.C.; Still, W.C. *J. Am. Chem. Soc.* **1989**, *111*, 4379–4386.

41. Goodman, J.M.; Still, W.C. *J. Comput. Chem.* **1991**, *12*, 1110–1117.

42. Holland, J.H. *Scientific American* **1992**, July, 44–50.

43. The line search is a procedure for evaluation of energy along a direction vector at several strategic points, depicting a cross section of the energy surface, defined by the intersection of the line search path and the energy surface. The line search procedure is costly as it requires three to ten functional evaluations per iteration to fully depict the cross section. However, when the potential energy surface is anharmonic, it is advantageous for convergence to fully extract all information from one direction vector before moving to the next. It can be used in conjunction with the most steep descent and full matrix Newton Raphson minimizations when the

starting geometries are highly strained or to improve or enable convergence on relatively flat potential energy surfaces.

44. Ponder, J.W.; Richards, F.M. *J. Comput. Chem.* **1987,** *8*, 1016.
45. Conjugate gradients requires convergence along each direction, as the gradient at i + 1 must be perpendicular to the gradient at i or the derivation of the mutually orthogonal gradients breaks down.
46. McCammon, J.A.; Gelin, J.A.; Karplus, M. *Nature* **1976**, *267*, 585.
47. Ryckaert, J.P.; Ciccotti, G.; Berenbsen, H.J.C. *J. Comput. Phys.* **1977**, *23*, 327.
48. Karplus, M.; Petsko, A. *Nature* **1990**, *347*, 631–639.

INDEX